# The Unravelers

# The Unravelers
## Mathematical Snapshots

edited by
Jean-François Dars
Annick Lesne
Anne Papillault

translated by Vivienne Méla

A K Peters, Ltd.
Wellesley, Massachusetts

Editorial, Sales, and Customer Service Office

A K Peters, Ltd.
888 Worcester Street, Suite 230
Wellesley, MA  02482
www.akpeters.com

Original edition:
Les déchiffreurs
© Éditions Belin - Paris, 2008

**Library of Congress Cataloging-in-Publication Data**

Déchiffreurs. English.
 The unravelers : mathematical snapshots / edited by Jean-Francois Dars, Annick Lesne, Anne Papillault.
    p. cm.
  Includes index.
  ISBN-13: 978-1-56881-441-4 (alk. paper)
 1. Mathematics--Miscellanea. 2. Mathematicians--Psychology. I. Dars, Jean François. II. Lesne, Annick.
III. Papillault, Anne. IV. Title.
  QA99.D4813 2008
  510--dc22

                                                                                        2008011415


Printed in India

12  11  10  09  08                                                    10 9 8 7 6 5 4 3 2

# Contents

# Prologue

The tides of fortune having washed us ashore one day in the Vallée de Chevreuse, on the startling banks of the Institut des Hautes Études Scientifiques (IHES), like the shipwrecked sailors in the travels of Sinbad, we were astounded by the customs of the inhabitants of this uncharted island. An instinctive reflex prompted us to take photographs of them, in their lonely offices, or attempting to scale the north face of the triptych boards of the lecture halls, chalk or pencil in hand, engrossed in dialogues, drinking with their eyes each other's words.

One of them, not the least among them, seeing these photos, thought that "if everyone wrote a little text" we could make a book out of it. To our wonder, everyone played the game. The texts arrived like autumn leaves, short, immense, violent, delicate, allusive, direct.... A whole human society with its essential fragility, composed of mathematicians, theoretical physicists, biologists, permanent residents or those just passing through, post grads or prize-winning researchers, tossed a multitude of little bottles into the sea. They are for you, reader, and for those like you, inhabitants of the lands we too come from, but who have not had the good fortune to approach these shores.

Jean-François Dars
Annick Lesne
Anne Papillault

Michael Atiyah
University of Edinburgh
Fields Medal
Abel Prize

# Dreams

In the broad light of day mathematicians check their equations and their proofs, leaving no stone unturned in their search for rigour. But, at night, under the full moon, they dream, they float among the stars and wonder at the miracle of the heavens. They are inspired. Without dreams there is no art, no mathematics, no life.

Michael Atiyah

The power of dreams: *such is the title given by Roger Caillois, the most rigorous of critics, to his famous anthology of oneiric literature. Day, night. Calculations, inspiration. Not opposed to one another but nurturing one another. During the day ordinary creatures work. At night, they sleep (sometimes it's the other way round). Those that Michel Atiyah speaks of, who will reveal themselves on page after page, constantly, tirelessly pass through the mirror, as one goes through a curtain, with the inimitable ease that can only be acquired through long years of effort.*

Alain Connes
Collège de France - IHES
Fields Medal
Crafoord Prize
CNRS Gold Medal

# Pitiless Reality

## 1. Foreword

This text describes a very personal relationship with mathematics, for let us not forget that each mathematician is "a particular case" and what is written here is valid only for the author and cannot be considered a "generic" point of view.

Mathematics is, to my mind, above all, the most elaborate tool of thought and generator of concepts that we possess in order to understand, in particular, the world around us. New concepts are engendered by a slow process of distillation in the alembic of thought.

It is tempting at first to divide mathematics into separate fields like geometry, the science of space, algebra, the art of manipulating symbols, analysis which gives access to the infinite and the continuous, number theory, etc., but this doesn't describe an essential feature of the mathematical world, that is to say that it is impossible to isolate one part without depriving it of its essence.

## 2. Act of Rebellion

In mathematics, in my view, the first thing to know is that one doesn't become a mathematician by learning; one becomes a mathematician by doing mathematics. So, it is not the "knowing" which counts, what is important it the know-how. Of course, knowledge is absolutely necessary—and there's no question of putting aside acquired knowledge—but I have always thought that one progresses more by stalling, faced with a problem in geometry, than by absorbing more and more ill-digested knowledge.

So for me, one starts to become a mathematician more or less through an act of rebellion. In what sense? In the sense that the future mathematician will start to think about a certain problem, and he will notice that, in fact, what he has read in the literature, what he has read in books, doesn't correspond to his personal vision of the problem. Naturally, that is very often the result of ignorance, but that is not important as long as his arguments are based on personal intuition and, of course, on proof. So it doesn't matter, because in this way he'll learn that in mathematics there is no supreme authority! A twelve-year-old pupil can very well oppose his teacher if he finds proof of what he argues, and that differentiates mathematics from the other disciplines, where the teacher can easily hide behind knowledge that the pupil doesn't have. A child of five can say to his father, "Daddy, there isn't any biggest number" and can be certain of it, not because he read it in a book but because he has found a proof in his mind.... There is freedom to be seized by him who recognizes it and respects the rules. And the first thing that counts is to become one's own authority. That is to say, to understand something, not to try and check immediately if it is written in a book; that's not the way. That will only delay his awakening to independence. What he must do, is to check in his mind if it's true. When one has understood that, one can little by little become very familiar with a tiny portion of mathematical territory and begin a long journey across these marvellous territories that one tries to reveal from one's own personal reference point.

## 3. Poetic Force

One can say that there are two aspects to the mathematician's task: the one which consists in proving, checking, etc. and which demands intense concentration, which requires exacerbated rationalism, but fortunately there is also the visionary aspect! And the visionary aspect is a bit like a setting in motion through intuition, which is not

subject to certitudes but is more like an attraction of a poetic nature. To simplify, there are two moments in mathematical discovery. There is the first moment when intuition cannot yet be formulated in transmissible terms in a rational manner. And at that point what counts is vision! Not the static aspect but a kind of poetic driving force.

This poetic force is almost impossible to put into words. When one tries to convey it, when one tries to say it, one only succeeds in turning it to stone, so to speak, and one loses the momentum which is essential in discovery.

Then, when enough pieces of the puzzle are in place and one can see that the vision can be translated into solutions of problems, things change. For example, when I started to become a mathematician, one of the things which struck me most in what I found—it was while writing my thesis with Jacques Dixmier—*(photo opposite)* was that a non-commutative algebra evolves with time! What I had shown was that in fact a non-commutative algebra has an evolution in time which is completely canonical. More precisely, the evolution given by the theory of Tomita, but which depended on a state, depended in fact only on this state modulo inner automorphisms, which are trivial, which don't exist. So that showed that it was non-commutativity which generated time! From nothing! Just like that! Of course the immediate result was that an algebra has many invariants, like, for example, its periods, that is to say the times $t$ when the evolution is trivial. But these results, although possible to formulate and transmit, do not exhaust the poetic content, the marvellous setting in motion of the initial discovery.

# 4. Mathematical Reality

There are poets that I admire very much, like Yves Bonnefoy, because of their proximity, on the methodological level, to mathematics. To my mind, what distinguishes the poet from the mathematician is that the poet's raw material is the physical reality of human experience. The main ingredient of poetry is the clash between the individual's inner world and external reality, the violence of which always takes us by surprise. On the other hand, the mathematician's journey is a trek through a different geographical space, through different landscapes, during which he comes up against another kind of reality. Mathematical reality is just as harsh, just as resistant as the physical reality in which we live. The moments of vision are not sufficient to enable the mathematician to actually do mathematics. That is to say that in counterpoint to the visionary aspect, in the moment which follows the proof, then come the hours of uncertainty, of suffering, the constant fear of having made a mistake. It's a bit like going down a cliff-face which forces one to constantly look down… You have to say to yourself all the time, "Here I could have made a mistake, perhaps I did make a mistake." You don't know and you're always scared! You may sometimes spend hours in terrible anguish, just because you have come up against true reality. So it's not reality in the ordinary sense, but it is probably even more pitiless.

The notion of truth is applied to another world, which is not the world of human experience in

external reality but that of mathematical reality. The crucial point to understand is that while so many mathematicians have spent their lives exploring this world, they all agree on its contours and its connectedness: whatever the origin of his itinerary, one day or another if the journey is long enough and if one refrains from confining one-self to an area of extreme specialization, one will finally reach one of the fabled cities, such as elliptic functions, modular forms, zeta functions, etc. "All roads lead to Rome," and the mathematical world is interconnected. Of course that doesn't mean that all its parts are alike, and Grothendieck in *Récoltes et Semailles* describes thus his path from

the landscape of analysis, where he began his journey, to that of algebraic geometry:

*"I still remember that striking impression (certainly quite subjective), as if I had left behind the harsh, arid steppes and was suddenly in a kind of 'promised land' whose luxuriant riches, multiplied to infinity, could be plucked, sifted wherever the hand chose to delve."*

Alexandre Grothendieck

## 5. Galois

What Galois understood to a certain extent, and what is almost the starting point of modern mathematics, is that in fact you must be able to go beyond calculation. That is to say, not to make calculations but to do them in your head! And to understand what their nature will be, to understand what difficulties you will be up against, etc., but without actually doing concrete calculations, to understand what form the result will take; what symmetry the result will have. And thus to go beyond the kind of outer envelope which you can easily become trapped in if you don't lift your eyes from the wheel. You have to try and rise above calculation, to meditate on the level of symmetries, etc.

*"To jump with both feet on the calculations; to group together operations, to classify them according to their difficulty and not according to their form; that, in my view is our mission."*

Evariste Galois

Whereas his predecessors looked for the symmetrical functions of the roots of an equation, Galois began by breaking the symmetry, to see what was going on.... His starting point is the arbitrary choice of a root's function, which does not allow for any symmetry. The miracle is that the group of invariance that he deduces passing from the function to the roots is in fact independent of the initial arbitrary choice.

Far from being outdated, Galois's ideas still irrigate contemporary mathematics, simply because of their simplicity and the movement they give rise to. Grothendieck's theory of motives is a natural generalization of Galois's theory in dimension

> 0, that is to say, if you like, to polynomials with several variables. These present-day developments, like those of Galois's differential theory, are placed directly in the dynamics of Galois's ideas. Here we should quote his letter and testament.

*"You know, dear Auguste, that these subjects are not the only ones I have explored. For some time my main meditations have been directed on the application to transcendental analysis of the theory of ambiguity. The aim was to see in a relation between quantities or transcendental functions, what exchanges we could make, what quantities could be substituted to the given quantities without the relation ceasing to take place. In that way we see immediately that many expressions that we might look for are impossible. But I don't have the time and my ideas are not yet developed enough in this vast field."*

Evariste Galois

## 6. Algebra and Music

In my view, it's crucial for a child to be exposed to music at an early age. I believe that exposing a child to music, at around five or six, tempers the preponderance in his intellect of the sense of sight, that incredible, purely visual gift, that a child develops very early on and which in fact is linked to geometry. Music enables us to balance that with algebra, that is to say music is linked to time exactly as algebra is. On the one hand, in mathematics there is a fundamental duality between geometry, which corresponds to the visual areas of the brain and which gives an instantaneous and immediate intuition. We see a geometrical figure and wham!

$$M^+ W^+_\mu W^-_\mu - \tfrac{1}{2} \partial_\nu Z^0_\mu \partial_\nu Z^0_\mu - \tfrac{1}{2c_w^2} M^2 Z^0_\mu Z^0_\mu - \tfrac{1}{2} \partial_\mu A_\nu \partial_\mu A_\nu - igc_w (\partial_\nu Z^0_\mu (W^+_\mu W^-_\nu -$$
$$W^+_\nu W^-_\mu) - Z^0_\nu (W^+_\mu \partial_\nu W^-_\mu - W^-_\mu \partial_\nu W^+_\mu) + Z^0_\mu (W^+_\nu \partial_\nu W^-_\mu - W^-_\nu \partial_\nu W^+_\mu)) -$$
$$igs_w (\partial_\nu A_\mu (W^+_\mu W^-_\nu - W^+_\nu W^-_\mu) - A_\nu (W^+_\mu \partial_\nu W^-_\mu - W^-_\mu \partial_\nu W^+_\mu) + A_\mu (W^+_\nu \partial_\nu W^-_\mu -$$
$$W^-_\nu \partial_\nu W^+_\mu)) - \tfrac{1}{2} g^2 W^+_\mu W^-_\mu W^+_\nu W^-_\nu + \tfrac{1}{2} g^2 W^+_\mu W^-_\nu W^+_\mu W^-_\nu + g^2 c_w^2 (Z^0_\mu W^+_\mu Z^0_\nu W^-_\nu -$$
$$Z^0_\mu Z^0_\mu W^+_\nu W^-_\nu) + g^2 s_w^2 (A_\mu W^+_\mu A_\nu W^-_\nu - A_\mu A_\mu W^+_\nu W^-_\nu) + g^2 s_w c_w (A_\mu Z^0_\nu (W^+_\mu W^-_\nu -$$
$$W^+_\nu W^-_\mu) - 2 A_\mu Z^0_\mu W^+_\nu W^-_\nu) - \tfrac{1}{2} \partial_\mu H \partial_\mu H - 2 M^2 \alpha_h H^2 - \partial_\mu \phi^+ \partial_\mu \phi^- - \tfrac{1}{2} \partial_\mu \phi^0 \partial_\mu \phi^0 -$$
$$\beta_h \left( \tfrac{2M^2}{g^2} + \tfrac{2M}{g} H + \tfrac{1}{2} (H^2 + \phi^0 \phi^0 + 2 \phi^+ \phi^-) \right) + \tfrac{2M^4}{g^2} \alpha_h -$$
$$g \alpha_h M (H^3 + H \phi^0 \phi^0 + 2 H \phi^+ \phi^-) -$$
$$\tfrac{1}{8} g^2 \alpha_h (H^4 + (\phi^0)^4 + 4(\phi^+ \phi^-)^2 + 4(\phi^0)^2 \phi^+ \phi^- + 4 H^2 \phi^+ \phi^- + 2(\phi^0)^2 H^2) -$$
$$g M W^+_\mu W^-_\mu H - \tfrac{1}{2} g \tfrac{M}{c_w^2} Z^0_\mu Z^0_\mu H -$$
$$\tfrac{1}{2} ig (W^+_\mu (\phi^0 \partial_\mu \phi^- - \phi^- \partial_\mu \phi^0) - W^-_\mu (\phi^0 \partial_\mu \phi^+ - \phi^+ \partial_\mu \phi^0)) +$$
$$\tfrac{1}{2} g (W^+_\mu (H \partial_\mu \phi^- - \phi^- \partial_\mu H) + W^-_\mu (H \partial_\mu \phi^+ - \phi^+ \partial_\mu H)) + \tfrac{1}{2} g \tfrac{1}{c_w} (Z^0_\mu (H \partial_\mu \phi^0 - \phi^0 \partial_\mu H) +$$
$$M \left( \tfrac{1}{c_w} Z^0_\mu \partial_\mu \phi^0 + W^+_\mu \partial_\mu \phi^- + W^-_\mu \partial_\mu \phi^+ \right) - ig \tfrac{s_w^2}{c_w} M Z^0_\mu (W^+_\mu \phi^- - W^-_\mu \phi^+) + igs_w M A_\mu (W^+_\mu \phi^- -$$
$$W^-_\mu \phi^+) - ig \tfrac{1 - 2c_w^2}{2c_w} Z^0_\mu (\phi^+ \partial_\mu \phi^- - \phi^- \partial_\mu \phi^+) + igs_w A_\mu (\phi^+ \partial_\mu \phi^- - \phi^- \partial_\mu \phi^+) -$$
$$\tfrac{1}{4} g^2 W^+_\mu W^-_\mu (H^2 + (\phi^0)^2 + 2 \phi^+ \phi^-) - \tfrac{1}{8} g^2 \tfrac{1}{c_w^2} Z^0_\mu Z^0_\mu (H^2 + (\phi^0)^2 + 2(2 c_w^2 - 1)^2 \phi^+ \phi^-) -$$
$$\tfrac{1}{2} g^2 \tfrac{s_w^2}{c_w} Z^0_\mu \phi^0 (W^+_\mu \phi^- + W^-_\mu \phi^+) - \tfrac{1}{2} ig^2 \tfrac{s_w^2}{c_w} Z^0_\mu H (W^+_\mu \phi^- - W^-_\mu \phi^+) + \tfrac{1}{2} g^2 s_w A_\mu \phi^0 (W^+_\mu \phi^- +$$
$$W^-_\mu \phi^+) + \tfrac{1}{2} ig^2 s_w A_\mu H (W^+_\mu \phi^- - W^-_\mu \phi^+) - g^2 \tfrac{s_w}{c_w} (2 c_w^2 - 1) Z^0_\mu A_\mu \phi^+ \phi^- -$$
$$g^2 s_w^2 A_\mu A_\mu \phi^+ \phi^- + \tfrac{1}{2} ig_s \lambda^a_{ij} (\bar{q}^\sigma_i \gamma^\mu q^\sigma_j) g^a_\mu - \bar{e}^\lambda (\gamma \partial + m^\lambda_e) e^\lambda - \bar{\nu}^\lambda (\gamma \partial + m^\lambda_\nu) \nu^\lambda - \bar{u}^\lambda_j (\gamma \partial +$$
$$m^\lambda_u) u^\lambda_j - \bar{d}^\lambda_j (\gamma \partial + m^\lambda_d) d^\lambda_j + igs_w A_\mu \left( -(\bar{e}^\lambda \gamma^\mu e^\lambda) + \tfrac{2}{3} (\bar{u}^\lambda_j \gamma^\mu u^\lambda_j) - \tfrac{1}{3} (\bar{d}^\lambda_j \gamma^\mu d^\lambda_j) \right) +$$
$$\tfrac{ig}{4c_w} Z^0_\mu \{ (\bar{\nu}^\lambda \gamma^\mu (1 + \gamma^5) \nu^\lambda) + (\bar{e}^\lambda \gamma^\mu (4 s_w^2 - 1 - \gamma^5) e^\lambda) + (\bar{d}^\lambda_j \gamma^\mu (\tfrac{4}{3} s_w^2 - 1 - \gamma^5) d^\lambda_j) +$$
$$(\bar{u}^\lambda_j \gamma^\mu (1 - \tfrac{8}{3} s_w^2 + \gamma^5) u^\lambda_j) \} + \tfrac{ig}{2\sqrt{2}} W^+_\mu \left( (\bar{\nu}^\lambda \gamma^\mu (1 + \gamma^5) U^{lep}_{\lambda\kappa} e^\kappa) + (\bar{u}^\lambda_j \gamma^\mu (1 + \gamma^5) C_{\lambda\kappa} d^\kappa_j) \right) +$$
$$\tfrac{ig}{2\sqrt{2}} W^-_\mu \left( (\bar{e}^\kappa U^{lep\dagger}_{\kappa\lambda} \gamma^\mu (1 + \gamma^5) \nu^\lambda) + (\bar{d}^\kappa_j C^\dagger_{\kappa\lambda} \gamma^\mu (1 + \gamma^5) u^\lambda_j) \right) +$$
$$\tfrac{ig}{2M\sqrt{2}} \phi^+ \left( -m^\kappa_e (\bar{\nu}^\lambda U^{lep}_{\lambda\kappa} (1 - \gamma^5) e^\kappa) + m^\lambda_\nu (\bar{\nu}^\lambda U^{lep}_{\lambda\kappa} (1 + \gamma^5) e^\kappa) \right) +$$
$$\tfrac{ig}{2M\sqrt{2}} \phi^- \left( m^\lambda_e (\bar{e}^\lambda U^{lep\dagger}_{\lambda\kappa} (1 + \gamma^5) \nu^\kappa) - m^\kappa_\nu (\bar{e}^\lambda U^{lep\dagger}_{\lambda\kappa} (1 - \gamma^5) \nu^\kappa) \right) - \tfrac{g}{2} \tfrac{m^\lambda_\nu}{M} H (\bar{\nu}^\lambda \nu^\lambda) -$$
$$\tfrac{g}{2} \tfrac{m^\lambda_e}{M} H (\bar{e}^\lambda e^\lambda) + \tfrac{ig}{2} \tfrac{m^\lambda_\nu}{M} \phi^0 (\bar{\nu}^\lambda \gamma^5 \nu^\lambda) - \tfrac{ig}{2} \tfrac{m^\lambda_e}{M} \phi^0 (\bar{e}^\lambda \gamma^5 e^\lambda) - \tfrac{1}{4} \bar{\nu}_\lambda M^R_{\lambda\kappa} (1 - \gamma_5) \hat{\nu}_\kappa -$$
$$\tfrac{1}{4} \bar{\nu}_\lambda M^R_{\lambda\kappa} (1 - \gamma_5) \hat{\nu}_\kappa + \tfrac{ig}{2M\sqrt{2}} \phi^+ \left( -m^\kappa_d (\bar{u}^\lambda_j C_{\lambda\kappa} (1 - \gamma^5) d^\kappa_j) + m^\lambda_u (\bar{u}^\lambda_j C_{\lambda\kappa} (1 + \gamma^5) d^\kappa_j) \right) +$$
$$\tfrac{ig}{2M\sqrt{2}} \phi^- \left( m^\lambda_d (\bar{d}^\lambda_j C^\dagger_{\lambda\kappa} (1 + \gamma^5) u^\kappa_j) - m^\kappa_u (\bar{d}^\lambda_j C^\dagger_{\lambda\kappa} (1 - \gamma^5) u^\kappa_j) \right) - \tfrac{g}{2} \tfrac{m^\lambda_u}{M} H (\bar{u}^\lambda_j u^\lambda_j) -$$
$$\tfrac{g}{2} \tfrac{m^\lambda_d}{M} H (\bar{d}^\lambda_j d^\lambda_j) + \tfrac{ig}{2} \tfrac{m^\lambda_u}{M} \phi^0 (\bar{u}^\lambda_j \gamma^5 u^\lambda_j) - \tfrac{ig}{2} \tfrac{m^\lambda_d}{M} \phi^0 (\bar{d}^\lambda_j \gamma^5 d^\lambda_j) + \bar{G}^a \partial^2 G^a + g_s f^{abc} \partial_\mu \bar{G}^a G^b g^c_\mu +$$
$$\bar{X}^+ (\partial^2 - M^2) X^+ + \bar{X}^- (\partial^2 - M^2) X^- + \bar{X}^0 (\partial^2 - \tfrac{M^2}{c_w^2}) X^0 + igc_w W^+_\mu (\partial_\mu \bar{X}^0 X^- -$$
$$\partial_\mu \bar{X}^+ X^0) + igs_w W^+_\mu (\partial_\mu \bar{Y} X^- - \partial_\mu \bar{X}^+ Y) + igc_w W^-_\mu (\partial_\mu \bar{X}^- X^0 -$$
$$\partial_\mu \bar{X}^0 X^+) + igs_w W^-_\mu (\partial_\mu \bar{X}^- Y - \partial_\mu \bar{Y} X^+) + igc_w Z^0_\mu (\partial_\mu \bar{X}^+ X^+ -$$
$$\partial_\mu \bar{X}^- X^-) + igs_w A_\mu (\partial_\mu \bar{X}^+ X^+ -$$
$$\partial_\mu \bar{X}^- X^-) - \tfrac{1}{2} g M \left( \bar{X}^+ X^+ H + \bar{X}^- X^- H + \tfrac{1}{c_w^2} \bar{X}^0 X^0 H \right) + \tfrac{1}{2} ig M \left( \bar{X}^+ X^0 \phi^+ - \bar{X}^- X^0 \phi^- \right) +$$
$$\tfrac{1}{2c_w} ig M \left( \bar{X}^0 X^- \phi^+ - \bar{X}^0 X^+ \phi^- \right) + ig M s_w \left( \bar{X}^0 X^- \phi^+ - \bar{X}^0 X^+ \phi^- \right) +$$
$$\tfrac{1}{2} ig M \left( \bar{X}^+ X^+ \phi^0 - \bar{X}^- X^- \phi^0 \right)$$

1

this way, and algebra. For example, I adore certain preludes by Chopin because I find that they have exactly the same marvellous property of condensation, of distillation. It's the kind of music which creeps into a room rather as if the window was suddenly opened by a gust of wind and then goes out the other side. To condense an idea in its most limpid, its purest form…that is, in some way, what algebra does.

# 7. Advice

I will end this text with a little practical advice:

### Go for a walk

When you are struggling with a very complicated problem (often involving calculations), the healthy thing to do is to go for a long walk (without paper or pencil) and to do the calculations mentally (ignoring the initial impression of "it's too complicated for that"). Even if you don't succeed, it's good memory training and it sharpens the teeth of the intellect.

### The divan

Mathematicians (male and female) generally have great difficulty in convincing their spouses that they work most intensely when they are lying in the dark on their beds. Unfortunately the invasion of computer screens and electronic mail has rendered this means of concentration less common: it is all the more precious for that.

That's it, that's all, we don't need to explain it, we don't want to explain it. On the other hand there is algebra. There is nothing visual about algebra; however there is temporality; it is "in time"! That's calculation, etc. It's something which evolves and it's something which is very close to language and which has the diabolical precision of language. And we can perceive that power, the elaboration of algebra, through music. So for me, there is an incredible collusion between music, perceived in

## Be courageous

There are two moments in mathematical discovery; there is a time when you have to be courageous: you have to climb up the rock-face and never look down…. Why? Because if you start looking down you're going to say: "Yes, but of course, So-and-So has already studied this problem and didn't succeed in solving it, so there is no reason why I should succeed." And you're going to find a hundred rational justifications to prevent you from moving upwards. So you must make an abstraction of that. You must "protect your innocence" in order for an idea to take shape without dissolving prematurely in the mists of knowledge at time $t$.

## Stress

During his lifetime (often right at the beginning) a mathematician is often confronted with difficulties due to severe competitiveness. For example, you receive a competitor's "preprint" on the same subject that you've been working on, and you feel you are under unreasonable pressure to publish quickly. The only recipe I know in these cases is to try and transform this feeling of frustration into energy to work even harder.

## Bad grace

One of my colleagues confided in me a long time ago: "We (mathematicians) work for the grudging approbation of a few friends." It's true that as research is work of a solitary nature, the researcher feels the need for approval in one way or another. The truth is that there is only one judge who really counts; that is you. And you can't bargain with that one. It is simply a waste of time to worry too much about the opinion of others; no theorem was ever proved by referendum, and as Feynman said: "Why do you care what other people think?"

Alain Connes

Dirk Kreimer
CNRS - IHES

# From Tasmania
# with Love

One hot summer day thirteen years ago in January 1994, I found myself sitting in my office at the University of Tasmania, Hobart, Australia.

Outside, I observed and ignored a bushfire approaching campus; inside, I was trying to make sense of the last email of a dear colleague, David Broadhurst, regarding some weird and wonderful numbers we had just found in the computation of Feynman diagrams.

I moved to Tasmania in 1993, tempted by the opportunity to visit Bob Delbourgo's group over there, and stayed on for a good two years. It turned out to be a perfect choice: theoretical particle physics, my home turf, is an area where large-scale computations are often the order of the day, and time for reflection is always short. Tasmania, I hoped, was a place to step back and rethink some theoretical foundations of particle physics.

It all started when we observed that certain amplitudes from topological simple terms in the Feynman graph expansion produce just rational numbers, whilst more intricate topologies gave much more distinguished numbers, periods of mixed motives, as we now have learned.

But to get to that understanding, data were needed, and that simply meant to compute more terms in that graphical expansion. Fortunately, David was a collaborator unsurpassed in skill when it came to such computations, so there I was, sitting in my office in Hobart—it had a window with a view indeed, with the Southern Ocean merely a mile away—scribbling down all these graphs, musing about their topology and exchanging emails frantically with David. It suited us well that he was just at the opposite point of the planet in the UK; his sleeping patterns and the ten hours difference in time zone found us online at the same time more often than not.

One of those emails suddenly enquired, "What's the bushfire doing, by the way?"—it had made it into BBC news, apparently. Well, I looked up and realized the physics building was deserted, and the eucalyptus trees some 50 yards away had started going up in flames, but that's also where it stopped. So our collaboration progressed happily, and numbers weird and wonderful we found enough. But those years in Tasmania, an island more densely populated by devils, spiders and sharks than humans, also laid the foundations for a much more profound understanding of the mathematical structure of quantum field theory and its perturbative expansion.

That quest brought me here to the IHES eventually, but that is another story, which would start again with a happy collaboration with Alain Connes. So the IHES is the place for me now to

reflect on quantum fields and how we use them in our description of the world. It's a place for reflection and thought in a spirit as solitary as those beaches along the Southern Ocean at the southern tip of that little island.

Dirk Kreimer

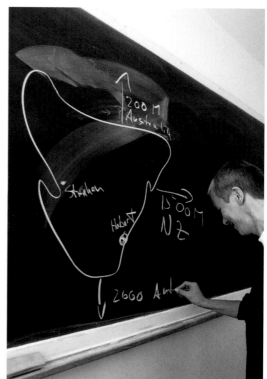

# Puzzles

I'm just in it for the puzzles. I like to play with interesting ideas, have fun turning them around, taking them apart, and putting them together. To be vague to the point of near meaninglessness, my tastes run to things which are recursive, which you can draw pictures to understand, which are built or grow out of each other. Complexity out of bits and pieces. Patterns and examples.

Pure mathematics is motivated both by the large—the big conjectures, the deep theorems, the grand visions; and the small—the puzzles, the little ideas you can hold in your hand. I tend towards the small.

I was very fortunate that Dirk Kreimer caught me in his net for lost souls. Dirk is someone who sheds new ideas whenever he moves. Now I can amuse myself with the thought that the puzzles I'm working on relate to the universe and other grand things after all. I was saddened recently with the realization of the utter vastness of the many tracts of mathematics full of interesting puzzles that I'll never have a chance to explore. I have a perhaps vain hope that one's doctorate is the narrowest part of one's career in this respect, earlier being wider on account of all possibilities being open, later on account of increasing expertise.

Karen Yeats

Paulo Almeida
University of Lisbon

# Structured Fury

"The very essence of mathematics is its freedom." Georg Cantor's famous saying thus conveys the intimate, if often unacknowledged, feeling shared by all mathematicians, whether they are disguised alternately as decipherers or creators. A freedom which is subject to reason alone, a concept which is not without contradiction since the mathematician voluntarily submits himself to rules of implacable rigidity in the free exercise of his activity. On the other hand, rationality, the touchstone both of modern science, which is moreover still Galilean, and of the democratic idea in its first stages in Greece, necessitates total freedom, which favours the systematic questioning of all supposed truths. For it is certain that one can have only the relative certainty of being certain. As it is for everyone, but perhaps more so for the mathematician, Freedom is at the same time an imperative and a condition of Reason.

The individual status of the mathematician anticipated the Renaissance idea that the use of perspective in art could be extended to the world at large, conferring on each one the liberating power of having his own *point of view*. On the other hand, however, the mathematician has never been able to free himself from the Roman sort of *status civitatis,* with its rights and contractual obligations according to a hierarchy of criteria; between the absolute of his identity and his belonging to a community, the mathematician comes to terms in his own way with Liberty and the Law; he creates and he reasons and doing this, mysteriously, he deciphers.

A thousand years of experience have brought us to believe that, in many areas of science, what is rational is real and what is real is rational, the singularity of mathematics being that reality is more often very far off, both in what concerns the origin of the ideas and its destiny—which in no way puts into question the mathematician's quiet, underground work. This work, with its erratic advances, its crazed anguish, sometimes a string of defeats, still reserves, here and there, the balm of secret, intimate pleasures, like the dog in Rabelais who chews a bone to suck out the *sustantificque mouelle* (the essential marrow), felt by the mathematician as a cog in the machinery which transcends him, particularly because of the repercussions on the real world. All this is not without apprehension, for the nature of mathematical ideas is always

21

the same, whether they are a source of benefit to humanity or of profit to the warmongers.

Imposing on themselves strict rules which encompass their liberty—essentially rules of argumentation—mathematicians have succeeded since the Greeks in discovering an original way of attaining truth different from the method of systematic experimentation introduced by the Renaissance: in both cases the ideas which are confirmed or infirmed being first glimpsed intuitively. The risks of damage caused by intuition are limited by proof or experimentation, according to the case, but it remains that it is intuition which triggers the process of discovery in both cases. And yet, in mathematics, the "private life" of

original ideas is not generally accessible, and one only knows in fact their "public life" in the formal appearance of definitions and proofs.

This circumstance induces a distortion with serious consequences in the teaching of mathematics since one is likely to believe that the dual life of mathematical ideas is reduced to their "public life" alone; but for the mathematician formalism, like the penal code, has a technical value which naturally has to be mastered; however, the real progress is founded on intuitions triggered by an instinct for the intelligibility of things, more specifically in mathematics by the instinct for what is rationally expressible; distinct by nature from the elusive rationality of the inexpressible that one finds in music or more generally in all the arts. There too a note of music, a dancer's gesture, a painter's brush-stroke, a writer's choice of words or a photographer's angle has its own indisputable necessity. However, if it is true that the elusive character of all this liberates us, at the same times it forbids us the elaborate and fragile constructions typical of mathematics, which remain pertinent whatever their distance from their place of origin.

Here we find again the opposition between the romantic explosion and classical rigour, both capable of excess, either by too much fury or by too much structure. Mathematical activity, forever searching for a balance between the two extremes, is nothing other than, to borrow an expression from a Portuguese philosopher Antonio Sérgio, a *structured fury*, but which is expressible.

Paulo Almeida

Ngô Bao Châu
Paris-Sud University
Institute for Advanced Study,
Princeton

# A Tartar Desert

Upon arriving at the IHES we ordinary mathematicians share the same feeling that Muslims experience on a pilgrimage to Mecca. Here is the place where, for a dozen or so years, Grothendieck relentlessly explained the holy word to his apostles. Of that saga, only the apocrypha reached us in the form of big, yellow, boring-looking books edited by Springer. These dozens of volumes entitled *Bois-Marie Algebraic Geometry Seminar* were the main staple of our apprenticeship years and are still our most precious working companions. Up there is the control tower where legend has it that Deligne shut himself up to find his proof of the Weil conjecture.

However, after a few weeks, the religious feeling passed and boredom set in. Nothing much happened in this pretty building surrounded by verdant countryside. Looking back, I am now sure that things were going on but without our noticing. Here is my personal experience to support this less than obvious conviction.

For some weeks I had been battling with an article by Faltings. He is undoubtedly one of the

Function-fields: $\amalg$-conj. $\overset{X}{\underset{B}{\pi}}$ rel curve of genus $g>1$ finite except for constant curves

strategy (Parshin, Arakelov): a) bound numerical invariants ($\deg \pi_*\omega_{X/B}$, $\omega^2_{X/B}$) parametrized by algebraic varieties

b) no deformations (rigidity)

a) Hodge-theory: $\Gamma(X, \omega^2_X) \hookrightarrow H^2(X, \mathbb{C})$ bounded dim, $\deg > 0$  weierstrass-points

b) rigidity: $\omega^2_{X/B} > 0$ $(H^1(X, \omega^{-1}_{X/B})=0)$,  Arakelov: $\#(\Lambda^g \pi_*\omega_{X/B}) \hookrightarrow \omega_{X/B}$ (even in char $p$ with maybe different exp)

char $p>0$: $S \geq p \geq 0$: $\kappa \neq 0 \to$ ok.

Assume we have $\infty$-many $g$ good approx $|z_i - \alpha| \leq \frac{c}{H(z_i)^\delta}$

Choose $z_1, z_2$ with $h(z_1) \gg 0$, $h(z_2)/h(z_1) \gg 0$

Choose $F(\pi_1, \pi_2)$ poly of bidegree $(d_1, d_2)$ $d_i \sim \frac{d}{h(z_i)}$

$0 \neq F(\alpha, \alpha) = 0$ to high order (weighted order) $\to F(z_1, z_2) = 0$ to high order

Argument that this is not possible (Roth's lemma)

Higher dimensions: $X$ variety, $x_1, \ldots, x_n \in X(K)$ rat'l points $d_i \sim \frac{d}{h(x_i)}$

$h(x_1) \gg 0$, $h(x_2)/h(x_1) \gg 0, \ldots$ $0 \neq F(T_1, \ldots, T_n)$ high order zero at $(x_1, \ldots, x_n)$

index $\delta$ if $\partial_1^{i_1} \cdots \partial_n^{i_n} f(x) = 0$, $\frac{i_1}{d_1} + \cdots + \frac{i_n}{d_n} \geq \delta$

$Z_\delta$: index $\geq \delta \subseteq X^n$ a $Z_0 \supseteq Z_{00}? \supseteq Z_{\delta} \supseteq \ldots$
common irred. comp $Z$

bound $\deg(Z_i)$, $h(Z_i)/(d/d_i)$ $x_i \in Z_i$

use induction. Vojta: $X$ ab variety clever choice of line-bundles

$Z \subset Y$, $Z(K)$ is finite if it does not contain a translate of an abel subvar.

Mordell-Weil $C/K$  number field $J(K)$

$\to J(K)$ is finitely generated

ingredients: $J(K)/2J(K) < \ldots$ finite

heights: $H(\frac{p}{q}) = \max|p|$

for function-fields: $\underset{s}{C} \hookrightarrow C \subseteq \mathbb{P}^n_K$ $k(B)$

$Sp(K) \subset B$ $h(s) = d$

curve $C(K)$ finite

only Vojta ($\sim 1990$) would use this.

$\dim = d$
degree $= \delta$, Chow $d, \delta$

connection-theory for number fields

$\to Z = \overline{Z}_K \subset C$ $h(Z) = \deg \overline{Z_i}$

Diophantine Approximation

$f(x,y) = 1$, $f$ hom. of degree $> 2$

say $f(x,y) = \prod_{i=1}^{d}(x - \alpha_i y) = 1$ $\alpha_i$ distinct

one factor small, others large $H(x,y)$

$(x - \alpha_i) \frac{1}{H(x,y)^{d-1}}$ $|\frac{x}{y} - \alpha_i| \leq \frac{c}{H(x,y)^d}$

Liouville bound $|\frac{x}{y} - \alpha_i| \geq \frac{c'}{H(x,y)^d}$

any better gives result

Thue, Siegel $\to$ Roth.

greatest geniuses among all practicing mathematicians, who, in addition, has the exquisite habit of writing his articles in such a way that one sometimes finds sentences which are completely meaningless. For a few weeks I had been trying to make sense of one of these sentences. In his own context it was possible to understand what he meant. However, his apparently absurd sentence suggested something in a wider context. It was that which interested me. That particular afternoon I had succeeded in giving it a precise meaning. Thereupon, though slightly disappointed after all my efforts, I had finally written out a lemma of ten lines that wasn't at all surprising, except to the extent that it seemed to have been unknown up till then. At teatime I told my lemma to Laurent Lafforgue, who replied with his usual enthusiasm: "But that's exactly what you have to do!" His enthusiasm warmed my heart without totally dissipating my doubts.

I now think that Lafforgue was right that afternoon; I had experienced one of the most decisive moments of my career.

Ngô Bao Châu

Paul-Olivier Dehaye
Merton College, Oxford

# Exoticism

"What is your research subject?"
"Why are you interested in it?"
"What use is it?"
"Why did that Russian refuse the medal?"
"What do you do from day to day, then?"
"Just paper and a pencil?"

Every mathematician is regularly confronted with a barrage of questions like these, testifying to the general public's fascination with his subject. The numerous books of vulgarisation which come out each year are another proof, as well as the multiplication of prizes and donations for the solving of the great problems of our time.

Where does this curiosity come from?

Perhaps it's the rather particular attitude of mathematicians. Or maybe the attraction of a truly exotic language. Or yet again, a hierarchical vision of the scientific disciplines which puts mathematicians at the top.

As for you, my friend, the reader, why are you browsing through this book?

Paul-Olivier Dehaye

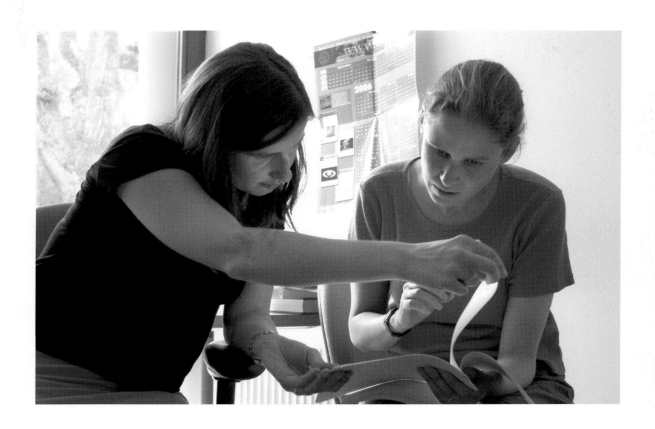

*It is not exactly the ascension of Mont Ventoux, but the solitary climber can rest assured that the audience will help him out in a difficult moment. Not necessarily through altruism: what counts is to reach the summit.*

*...ator is in trouble, he stumbles suddenly on an unexpected pebble, what can be done?*
*...e zero? No. A friend, a fellow, a colleague hurries forward with a fresh look, a new*

It's a game two can play: it requires an enthusiastic and a reticent player. On _____
because he knows that he knows, the other "reticises" until he has exhausted all pos

*However, it is not a confrontation, but a common effort towards a goal so subtle that it can only be reached at the cost of implacable rigour.*

*A strange area where only verified truth counts, where not to make any favours is the greatest favour.*

*Finally, sometimes, between two great masters (like at chess) the spectacle changes: while one unfolds the motives of the sumptuous Persian carpet that he has woven, the other, a connoisseur in*

Sophie De Buyl
Free University of Brussels -
IHES

# Curiosity

It appears that sixty-five billion particles (interacting very weakly) called neutrinos pass through every square centimetre of our skin every second, that time passes more slowly on the banks of the Mississippi than at the summit of Everest, that our body is mainly empty, that the speed of light is finite....

I am fascinated by man's capacity for abstraction and deduction. Through reasoning, which is sometimes quite simple, he succeeds in deciphering the world around him and understanding it beyond what he can apprehend through his five senses. His insatiable curiosity is a source of more and more questions, and the adventure will probably be without end. We may easily wager that a considerable number of fundamental interrogations as yet imprecise which torment the physics community today, such as the exact nature of space and time, will one day be perceived in a totally different light, a hundred years or so from now. Just as we know today that the Earth turns round the Sun and why the sky is blue.

What's more, did you know that our universe might have ten time-space dimensions, that there might be a black hole at the heart of out galaxy, that we might be living on a "membrane", like an ant on a string, only conscious of a part of the space in which it is plunged, or that each elementary particle might possess a "supersymmetric" partner?

The principles of symmetry have led, among others, to the standard model and general relativity, the two pillars of present-day physics. We can suppose that these principles will guide future progress in theoretical physics. I like it when the answers found involve elegant mathematics, when they unify concepts and give us the feeling of having reached a deeper understanding of the world around us.

Sophie De Buyl

Thibault Damour
IHES
Einstein Medal
Powell Medal

# The Paving Stones of the Route de Chartres and the Polynomials of Jones

It was on stepping onto the uneven paving of the courtyard of the Hôtel de Guermantes that the Narrator experienced the joy of Time Refound. I had a similar experience when, one bright autumn day in 1989, as a newly appointed "full professor", I crossed the paving of the gateway to 35 route de Chartres and saw the marble plaque with its shiny golden lettering: Institut des Hautes Études Scientifiques.

All of a sudden I saw myself fifteen years earlier and experienced again the moment when, chaperoned by Achille Papapetrou, I crossed the threshold of the IHES for the first time. I was coming to the seminar of John Archibald Wheeler. I can see again the glimpses of the park in spring through the bay windows of the "music pavilion", where the seminar took place and the fluttering of the large sheets of paper on which J. A. Wheeler had written and drawn in advance the subject of his talk with different coloured felt-tips. This seminar was an important moment in more than one way for the young man of twenty-three that I then was. First of all because of this impressive encounter with the ideas of a great physicist and with the man himself. Also because

of the intuitive understanding (for the inexperienced research student that I was) that it was better to avoid the programme of research, profound but far removed from observable reality, in which Wheeler was henceforth engaged (quantic gravitation in "super space"). Finally, because as a young graduate of the "École Normale", benefiting from a Jane Eliza Proctor postdoctoral grant from Princeton University, I had come to ask Wheeler if he would accept me for the academic year 1974–1975 in his group specialised in relativist gravitation. He accepted, and my pos-

tdoctoral stay (1974–1976) in Princeton played a crucial role in my scientific development and in my career.

Another fade in, fade out. A fifteen-year leap forward. The newly elected full professor that I now am takes his first lunch in the IHES cafeteria. Intimidated by my ignorance, particularly in mathematics, I feel out of place among the guests with famous names. On my left is Vaughan Jones whose famous "polynomials" have recently become the talk of the scientific town, following an article by Edward Witten showing the links

with certain models of physics. Will I dare admit to him my ignorance on "polynomials"? A jovial man, in both manner and appearance, Vaughan Jones kindly puts me at ease and offers to enlighten me. Using the small sheets of paper and pens always at hand on the refectory tables, he explains to me, in a few words and symbols, the role, the essence and the definition of Jones's polynomials. I then understood two things: first of all that the particular structure of the IHES favoured free and friendly exchanges (at lunch, at tea, before the blackboard…) between scientists from various fields, all of the highest level, thus affording a privileged access to knowledge that it would have been much harder and less efficient to acquire through reading published texts. Secondly, that I was so lucky to have been chosen to belong to this heart-warming scientific Thebaid that I would have to surpass myself to be worthy of this privilege and that I would be very happy at the IHES on both the intellectual and personal level.

Thibault Damour

*One is allowed to linger a little at the Guermantes'. Among all the algorithms created by Proust, there is one thanks to which a gesture, a word may bring to life the past of a place, the metamorphoses which have taken place before it becomes, for a moment, our reality.*

*Cécile DeWitt created the Summer School at les Houches, Yvonne Choquet-Bruhat was the first woman elected to the Academy of Science. Meeting them in the lounge at teatime, as one might greet the Duchess walking in the Bois de Boulogne, suddenly fills the room with the heraldry of modern physics.*

Cécile DeWitt
University of Texas

# From 1948 to the Present Day

I met Léon Motchane for the first time on November 13th 1948 at Princeton; I was then a post doc student at the Institute for Advanced Study. He had asked me to make an appointment for him with J. Robert Oppenheimer. It was a simple thing and I would quickly have forgotten all about it if Freeman J. Dyson, also a post doc student, hadn't described Motchane's visit in his weekly letter to his parents. Dyson's letters to his parents and, after their death, the letters he wrote to his sister contain marvelous tales, some of them published, of the many interesting encounters he made during his life. On the 14th of November, Dyson wrote to his parents, "Cécile amused us yesterday by bringing a French millionaire (a kind of industrial tycoon) to show him the institute. She told us she had suggested that an institute of this kind would be very useful in France; she told us that she had been nominated head of this French institute and that we would all be invited to teach there. It will be interesting to see what comes of this idea." [1]

Once I got back to France, my aim was to create a Summer School in Theoretical Physics. Motchane's aim was more ambitious: to create an institute on the Princeton model. But our desires were of the same order. So I went to see him. The Summer School could be a second home for his Institute. Our discussions were very cordial. We were the only people to set up new institutes in this field. But I was in more of a hurry than Motchane. I had fixed the creation of the Summer School as a condition to my marriage with a foreigner. Furthermore Motchane was looking for subsidies in the private sector; I was looking everywhere; I ended up with a subsidy from the Direction of Higher Education at the Ministry of Education. On April 18th 1951 the Council of the University of Grenoble created the Summer School of Theoretical Physics at Les Houches (Haute Savoie). On April 26th 1961 I married Bryce DeWitt. In 1958 the IHES was created (at the Thiers Foundation in Paris). In 1962 the Institute moved to Bois-Marie at Bures-sur-Yvette.

For 50 years I have frequented the IHES and participated in its activities in various capacities, but always with the same enchantment.

A few landmarks: in the 60s during a cocktail party at the IHES I met Pierre Cartier and disco-

vered his competence in the most varied fields. In 2006 our book *Functional Integration, Action and Symmetries* was published (Cambridge University Press), the fruit of twenty years of collaboration. This work would never have been accomplished without the marvelous working conditions at the IHES: the possibility of staying there for a longer or shorter period according to my other commitments; the incomparable organization (I can make this comparison because I have traveled a lot)—living conditions, meals, offices, competent staff always happy to see to the needs of others; the spirit of the place, le Bois-Marie, where one comes to life again.

On November 17th 1999, on the occasion of a "friendship day" Jacques Friedel presented me with the medal of the "Légion d'Honneur."

Since 1996 I have taken part in the life of the Institute as a member of the Administrative Council. Every meeting gives full opportunity to measure the progress accomplished under Jean-Pierre Bourguignon's direction. When one looks back at the distance covered since 1948, two names dominate the history of the IHES: Léon Motchane and Jean-Pierre Bourguignon: the first created it, the second has given it an impulsion which guarantees its permanence. Both of them have put their whole hearts into the task, and many are those who owe them gratitude.

Cécile DeWitt

[1] *Physics Today*, 42 (2) : 37, Feb. 1989.

Yvonne Choquet-Bruhat
Pierre and Marie Curie University
CNRS Silver Medal

# Knowing, Understanding, Discovering

A great mystery which presents itself to each one of us is that we are conscious of two distinct entities—our thought and the outside world. "I think, therefore I am," said Descartes. However, all of us (or nearly all of us) admit that "the other" also exists. A reality that each of us yearns, in one way or another, to know, in order to master it or simply to understand it. But what does it mean, "to understand"? This word has many meanings. For the scientific mind it means first of all to classify phenomena and pinpoint the relationships between them. The classification and the relationships exist. This is an observable fact that distinguishes outside reality from that which we construct sometimes in our sleep. The second stage is to group these relationships in a more general law of which they are the consequences and, finally, to transpose this law in a model constructed by our thought. Perhaps to understand is always to find the adequacy between our thought and a part of the reality of the outside world.

In scientists' construction of models, mathematics is a fundamental tool. It has long been useful; it is now indispensable for the formulation of observed facts. Reality has been revealed to be much richer and stranger than what our senses can perceive. It is difficult for a physicist to describe with exactitude the quantic phenomena using everyday words, but ever more elaborate mathematical models can translate their properties…. For the mathematician, these models become reality itself. It is marvellous to see that mathematics, a tool of thought and generator of concepts, possesses such an adequacy with reality. However, I think that no model can completely exhaust reality. I hope that new tools and observable facts that we cannot foresee will reserve surprises for future generations. The extraordinary complexity of biological systems now also requires mathematical modelization. This in turn initiates the creation of new branches and attempts to construct a model of thought itself. However, will this be enough to explain the "I am" of Descartes's aphorism?

Having expressed these reserves on the capacity of mathematics to find a final solution to every-

47

thing, I shall now sing its praises. It is a universal language where truths are absolute and unquestionable even if their verification is sometimes arduous. With any other language it is difficult to convey completely a relatively subtle thought, even with one's mother tongue, let alone to translate it in another language. Mathematics is a precious tool to construct models of experimental realities but it is also a marvellous reality in itself, which all intelligences on planet Earth (and maybe on other planets) can consult if it interests them. Many physicists searching for models of physical phenomena have discovered new mathematical concepts which have fascinated them. The mathematical beings thus created have lives of their own and have engendered others. A lot has been written about the mutual fertilization of mathematics and physics, and I will say no more about it.

Let me move on to more personal considerations. For me, working in mathematics is an escape route into an ideal world, where the journey is limited only by one's own limitations. The country is full of truths to be learned or truths to be discovered; I love to learn, but it is fantastic to find a new truth, even a very small one. A mathematical result concerning a model coming from physics has a very special flavour, because it presages a yet unknown property of the inexhaustible reality we are immersed in. How pleased the researcher is, even if the discovery is a very minor one! I love the mathematician's trade, a mixture of reasoned vision and the skilled work of calculations. One cannot make calculations without a guiding thread concerning their structure and aim, neither with the human mind

nor the computer. Certain results—sometimes surprising—are obtained only after long calculations which one cannot always circumvent in later proofs.

I will end by saying that the mathematical universe exists through the community of mathematicians who create it—or who discover it, if the reader prefers that philosophy. It is a great joy for a mathematician to belong to this community of citizens of the same ideal country. Specialists of the same discipline, whatever their nationality, share a certain number of truths and curiosity for unresolved problems. Their common knowledge and interests unite them more than rivalries of priority could divide them. Exchanges of points of view are stimulating and enriching. Work in collaboration is particularly gratifying. Mathematical affinities sometimes turn into real friendship, the salt of life.

Yvonne Choquet-Bruhat

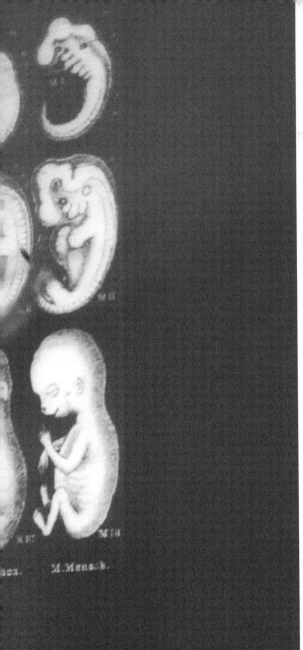

Arndt Benecke
CNRS - IHES

# Farewell to Alexander's Sword

As to the young mind a seeming contradiction between *mathematical logic* and *painless learning at school* might exist, as to the spiritual mind a seeming contradiction between *physical causality* and *freedom of choice* might exist, as to the productivity-oriented mind a seeming contradiction between *theoretical abstraction* and *getting things done* might exist, in the same way a seeming contradiction might be perceived to exist between a trained *biochemist at work* and a *pure mathematics and theoretical physics institute.* Here is the easy way out: "Physics has created powerful theories on objects and their interactions; mathematics is a formal language of description of objects and their interactions; biochemistry is the study of physical objects relevant to life and their interactions and therefore can/should/does rely on physical theories and mathematical descriptions." In other words: "*Painlessly learning* and using *freedom of choice,* the guy is *getting things done* [or trying to] with respect to some biological problem."

Despite the fact that this approach is my official current bread and butter, I like to think of the underlying logic as what the sword of Alexander the Great is to the Gordian knot; it is unidirectional (sword cuts knot, however, the knot does not do anything to the sword) and therefore axiomatic (the *functional form* of the sword is not derivable from the *topology* of the Gordian knot).

So what if Alexander's sword was too blunt to cut the knot? What if all descriptions of emergent biological *functional forms* founded on axiomatic set theories did not reduce the *complexity of description* and therefore fail to *abstract* the very nature of the object under study? What if the equivalent of the sword for emergence in biology still needs to be created rather than only applied? What if we actually could add to theoretical physics and mathematics by studying biological emergence? What if studying functional biological forms was a more powerful inspiration to formalism than studying inanimate objects, as the former display much higher levels of emergent organization and functionality (one might add beauty)?

Just kidding! The Gordian knot (under the assumption that the myth did not invent a "knot" not encompassed by our topological definition of a knot) ceases to exist in four dimensions. But then again, what if…maybe I am not kidding after all.

This institute simply stimulates and preserves the intellectual freedom a scientist requires to explore the creativity of her mind. That is why, at least in my mind, there is no contradiction in my being here.

Remark: A much more beautiful and elegant way of expressing some of the above ideas has been found, for instance by Douglas Hofstadter in *Gödel, Escher, Bach: An Eternal Golden Braid,* and also transpires from the works on morphodynamics of René Thom, whose former office I am deeply honored to have inherited.

Arndt Benecke

Time has long passed since Lamartine, addressing the world, asked if inanimate objects had a soul. He was fond of pointedly asking himself pointless questions as long as they were decorative. Nowadays we calmly reverse the proposition, twisting it a little: animate objects (you, me and everyone we know, cyanobacteria, Asian elephants), can you be measured, described, classified in the same way as your inanimate cousins? We need tools more refined than Alexander's sword: certainly we can gain time by slicing through the Gordian knot, but we still haven't unraveled it. And anyway, we're using the wrong scales.

Annick Lesne
CNRS - IHES

# Dialogue on the Scales of Life

*Question – To apply mathematics to biology—isn't that a rather banal idea?*

Annick Lesne – Indeed the idea isn't new: to use mathematics to analyse data on living organisms, to conceive models, to structure reasoning, lots of people think of doing that. However, to do it the other way round, starting with the living to take a fresh look at mathematics, is less usual.

*Q – But what is "life" for a mathematician? How am I different from a stone?*

A.L. – Because you can reproduce (proving in that way that the "viable solution" that you represent triumphed because it was able to reproduce itself quicker than all its possible variants). And so you are the product of innumerable histories, whereas the stone, which certainly is deformed over time, breaks in half, is rounded, becomes a pebble, etc., is the product only of its own particular history. As a living being (cell, human being, ecosystem), the product of past interactions, you are the bearer of a mine of information, simply by the fact that we can see you today!

*Q – Fantastic! It's a scheme which works for everything. Take three lines in a newspaper; by successive deductions you can work back to the whole of the history of the world. Any concrete fact is an end product.*

A.L. – Yes, but in a biological structure, there is an additional complexity: the scales of life. A living organism is composed of several levels which do not "know" each other, which don't "understand" each other directly but which must nevertheless "get along". The logic of proteins is not the same as that of cells, which again is not that of tissues, which is not that of the organism as a whole. And yet the whole is coherent!

*Q – So you would like to find the mathematical tools which would enable you to understand that coherence?*

A.L. – Yes, but not only that. Yes: in the sense that we are looking for the equivalent for living organisms of auto-similarity for fractals. If you try to describe a fractal by assigning a position to each point, then it's infinitely complicated. From the

perimental observations were taken up by some experiments in physics, then theorized by Einstein, validated by the experiments of Jean Perrin, and finally formalized in the framework of mathematics by Norbert Wiener. At this stage there was still some interaction and the Wiener process is still used by physicists, but it has spread into a whole host of mathematical processes which have totally lost sight of their origin in physics! And it's a mathematical theory which won Wendelin Werner the Fields medal.

*Q – What a lovely story! Except that you're not there yet...*

A.L. – Indeed we are a long way off... But I seriously believe that by studying groups of transformations closely we should be able to give evidence for types of groups of invariance and above all the failures of invariance which characterize life, even if this idea irritates some biologists, for whom life is reduced to a succession of events. A succession of miracles in fact, but that doesn't mean there isn't a certain mathematical logic behind it.

*Q – I didn't know that the notion of miracle was acceptable in your disciplines?*

A.L. – And what if I say: a succession of singular events with almost null probability among all the events permitted by the laws of physics? That would give us quite a good definition of life.

Annick Lesne

moment you say, "It's auto-similar and if I zoom in the picture, I find the same thing", then you can write the programme describing a fractal in two lines! Not only: in the sense that the mathematical tools which will resolve these questions on the multi-scale coherence of the organisation of life, will, because of their particularities and specificities, bring with them new questions. Which will induce new mathematics.

*Q – Which will distance itself from biology...*

A.L. – So much the better. Let's make a parallel with Brownian movement, at the frontier between maths and physics this time. Everything started with Robert Brown, a Scottish botanist. His ex-

*The main thing is to agree on the words. Mathematicians as well as theoretical physicists are extremely sensitive to the meaning of words. Just watch them when you use a term which for them has a precise professional meaning and which the rest of humanity uses in a wider acceptation. They balk. It's not that they are offended, no, it's because they just don't understand. Something gets stuck, you have introduced in the reasoning a link which has nothing to do with the others. You try to clarify your idea, they try along with you, they suggest a term which seems appropriate to them, you agree, their faces light up, the discussion can continue, running smoothly like an engine after a temporary problem with the carburettor. That's why it's so delightful to tackle them at their own game.*

These exchanges generally take place at teatime in the lounge of the Institute. A sacred and unwritten rule, in this universe without obligations or sanctions but where intellectual concentration is maintained at boiling point, decrees that everyone should meet up at least once a day to sip this delicate nectar. While some immerse themselves in the daily papers, others spar with chalk, as one spars with a sword, on the little blackboard in the corner (the benevolent administration has scattered boards along the corridors at strategic points) or search for the right word with uninitiated visitors. It is they who are right, for the strength of their endeavour does not allow for improvisation, and the solidity of the whole project depends on the correctness of the details. Mathematics requires such profound self-investment that it governs even the meaning one gives to one's life.

Nikita Nekrasov
IHES
Hermann Weyl Prize
Jacques Herbrand Prize

# Lost in Translation

For me the IHES represents a point of singularity. It combines all that I dreamed of since I was a child. Moreover, it combines it with what I kept dreaming of when I grew up. It is located in France. It is located in the country yet it is near Paris. It hosts both physicists and mathematicians, a good deal of whom work on the topics I find interesting to talk about. In addition, these physicists and mathematicians come from the countries which gave me an education and the taste for science. It is an unfortunate fact of life that these countries, in a sense, do not exist anymore. The Soviet Union has collapsed. The United States had 9/11. People seem to care about things they didn't care about before. And yet, the visitors at the IHES seem to live in a different world, where nothing but science is worth thinking about.

Lev Landau once said that all theoretical physicists come to physics from mathematics. I got interested in theoretical physics for the "wrong reason", and my love of mathematics came along, just as the love of music or poetry, or even of dancing, can come to a fellow who is shamelessly using them to court a girl of his fancy. At the age ten or so I was very fond of doing things with my hands, designing some little radio gadgets and electric car models (none of which worked too well, I must admit). I was interested in theoretical things even at that time, mostly the French language and French history (which was indeed a theoretical concept in the Soviet Union in the early eighties). In fact, my liaison with France started when I was five years old. My grandmother taught me some French, which she managed to remember from her pre-revolutionary school days

I got admitted to a school where French was taught at the age of seven or eight years old (in most Russian schools foreign languages are taught starting at the fifth grade, e.g., at the age of 12–13).

When I turned eleven, my parents moved; I changed schools, and instead of the French language and some notion of French culture, I got to study English as a foreign language, and had the ordinary street-smart hooligans as classmates. Luckily for me, I also had a great physics teacher, who recognized my genuine interest in the subject and started giving me advanced physics lectures.

He also pushed me towards theoretical physics and towards mathematics. Eventually I got into a high school with mathematical classes, so that the last three years of my school education were spent in the "right" environment. One aspect of our education was the absence of lectures on mathematics (except for elementary geometry). Instead, we learned set theory, calculus, and the introduction to differential geometry by solving mathematical problems our teacher would type for us. As far as physics was concerned I was left to myself; I was mostly interested in astrophysics, notably the evolution of stars, and tried to write a computer program which would describe a star's life and death (of course, I wanted to see a supernova turning into a black hole). As this problem turned out to be too complicated for me, I decided to "work" on something more general, like the evolution of the universe as a whole. One thing led to another and eventually I came across a paper by Michael Green

in *Scientific American*, called "Superstring theory", which had beautiful pictures of the Riemann surfaces (of which I knew nothing of course), and fell in love with the subject. From that moment on (I was fourteen years old) I knew what I wanted to do. That determined my choice of university, graduate school and so on. Of course, I had no idea of moving to another country.

I was probably a first year undergraduate student when I first learned about the IHES. I remember

vividly that I bought a book by Misha Gromov, *Differential Relations with Partial Derivatives*, which was translated into Russian, and read it with my friends who studied mathematics. We liked very much the "h-principle", which seemed like a very powerful method of proving various theorems. In fact, we liked it so much we even invented a special name for it, pronouncing the letter "h" in a funny way. I had a feeling that this book might be important for me as a future string theorist, but it

was a bit far from what I thought a string theorist must need first. Since the appeal of string theory is the quantum theory of gravity, I needed a good book that would help to quantize gravity, and of course four-dimensional gravity seemed like the most important thing to quantize. Remarkably, the book store I would go to while walking my dog had a book with a promising title, *Four-Dimensional Riemannian Geometry,* which was a collection of Arthur Besse's seminar proceedings.

This is where I learned about the great French school of differential geometry (algebraic geometry would come next) and first "met" my future Director, Jean-Pierre Bourguignon. Unfortunately, this book didn't provide me with all the vital information about four-manifolds I needed, and I didn't succeed in quantizing the four-dimensional gravity right away. But I did learn that the participants of Besse's seminar were very much interested in some manifold that was called K3. That I could

definitely relate to, since my father, in his youth, was a mountaineer, and our bookshelves were full of books on Annapurna, Chomolungma, K2 and so on. I was a bit perplexed as to why mathematicians were to study Kangchenjunga (the third highest mountain in the Himalaya, after Everest and K2) but of course, eventually, I learned that K3 is neither the mountain, nor the nuclear submarine, but rather the name that hides three great Ks (Kahler, Kodaira, and Kummer).

My encounter with the physics part of the IHES was more dramatic. As an undergraduate interested in such an abstract and "un-physical" subject as string theory, I had to prove first that I was capable of doing the real physics, e.g., phenomenology of weak interactions. The first thing one learns in this course is muon decay. The textbook I used (by Lev Okun) had the general formula for the lifetime of muon, without assuming the particular form of the four-Fermi interaction. The parameter entering this formula is called Michel parameter. Of course, I remembered from my childhood that Michel was a French name, so I knew it had something to do with French physics. I learned later about Louis Michel, and his role in French science and society.

A few years later I crossed the Atlantic and started my graduate studies. One day, at lunch, I was introduced to a cosmologist who looked distinctly European. Thibault Damour, as I found out, was working with Sasha Polyakov on a topic many people thought to be too courageous— they were trying to find testable predictions of string theory. I must say that I am not very proud of my behavior on our first encounter. Instead of discussing the grounds for their "minimal couplings principle," I launched a politically incorrect discussion of French girls, purely theoretical of course. I felt bad about it for some time, until one day I had lunch with my former adviser, Lev Borisovich Okun, who told me about his lunch with Bruno Pontecorvo in some open café in Marseille. Pontecorvo was in his late seventies, and suffered from Parkinson's disease. It was such a windy day that Lev Borisovich struggled to keep his napkin on the table. At this dramatic moment Pontecorvo thoughtfully addressed him: "Did you notice, Lev Borisovich, how much nicer the girls of Marseille are, compared to the Parisian ones?"

While I was studying in graduate school, Krzysztof Gawedzki invited me to take part in the program he organized in Vienna at the Erwin Schrödinger Institute. I went there and took advantage of the quiet atmosphere of Vienna cafes to write up my thesis. After that I got my first invitation to visit the IHES. Unfortunately I had to defend my thesis at that time and I couldn't come.

I did come to the IHES once, unofficially, while visiting my collaborator, friend, and in many respects, teacher, Samson Shatashvili, who was and, I hope, will be a regular visitor at the IHES. It was in the summer, several visitors were teaming up, headed by Maxim Kontsevitch, to play volleyball in the residence de l'Ormaille. I couldn't resist the appeal of the game, and of course got a good deal of criticism from my collaborator (a tennis champion).

A couple of years later I received another invitation to a program, organized by Mike Douglas. That time I had a family emergency and couldn't come either. Apparently, the suspense got unbearable, and the IHES made me the offer of a permanent post of professor. That offer, I couldn't refuse.

A few years later I learned that the main reason for hiring me was the fact that I played volleyball.

At some point I read a paper by Sasha Polyakov, the hero of every string theorist, either admired or envied. He concluded by saying that the IHES is the best place in the world to do physics. As always, he skipped some of his intermediate calculations, so that the result, which is undoubtedly correct, is very hard to reproduce. I keep working on it, at the risk of being fired for not playing enough volleyball anymore.

Nikita Nekrasov

Yiannis Vlassopoulos
University of Athens

# Thought Technology: The Pursuit of Structure

Shrouded in a certain amount of mystery, mathematics is a source of both some repulsion and a measure of fascination for most people these days. Nonetheless "doing mathematics" is doubtlessly one of the basic abilities of the human species and one of the major ways to make sense of the world around us. If we consider culture and civilization as an open source code (software), inherited and enriched from generation to generation, then mathematics could be one of the most universal parts of the code. A testament to this is the fact that mathematicians from all different backgrounds, cultures and nationalities have absolutely no problem communicating and collaborating when it comes to their work.

Mathematics is very often identified with logic, and certainly for something to be considered part of mathematics there has to be a proof for it, e.g., a sequence of logical steps starting from some basic definitions and leading to it. Nevertheless most working mathematicians would agree that "insight" strengthened by experimentation (working out examples) and aesthetics is the way to arrive at a guess of what should be true. In a way,

that putative truth is a vision of a destination while the proof has a dual role: first, that of verifying that the destination really exists, and second as a "highway" (or even a windy path) and a set of basic instructions for how to get there. If we elaborate every instruction down to its most elementary form, then we have a sequence of logical steps that any person can follow. Following each individual step, though, doesn't guarantee understanding of the journey or the destination. In other words, following one instruction after the other doesn't mean that one can create a mental image and assimilate it with the rest of the "software" already in use in one's brain. On the other hand, for someone who understands, building a highway and/or driving along it may offer a view and a way to destinations other than the original one, and that is also the case with proofs.

One may say that the essence of mathematics is to identify, distill and analyze structures. The presence of structures is usually interpreted in the form of operations, meaning that we are able to construct objects along with a procedure such that, given one, two or more of the objects, we can, by applying the procedure, produce more objects of the same kind. Everybody who knows how to add and multiply numbers is of course familiar with this, but we should point out that there are many different kinds of objects that arise in trying to encode structures. We will see, for example, a situation later on that involves operations between surfaces.

Admittedly, this way of presenting things seems to describe mostly what we call algebra, but there is a well known interplay between geometry and algebra. Geometric problems seem *a priori* more amenable to an intuitive approach, but it is usually a great advancement when an algebraic model is constructed that captures the situation.

There is, though, a true problem of exposing mathematics to the non-mathematician. I believe the problem is similar to what musicians would have if they were not able to play their music, but rather one had to read the notes from a piece of paper in order to experience the music. It wouldn't pose too much of a problem to musicians, but it would to other people. In other words communication would be much simpler if we could find a way to "play mathematics" just like we play music. To "play a piece of mathematics" one could use its application to a physical situation. For example, the meaning of the heat equation:

$$\frac{\partial}{\partial t} f(x,t) = \frac{1}{2} \frac{\partial^2}{\partial x^2} f(x,t)$$

can be illustrated by heating up a point on an iron rod and observing the distribution $f(x,t)$ of heat on the rod (at a distance $x$ from the point, after time $t$). It can also be demonstrated as a special case of the Black-Scholes equation in terms of

the price $f(x,t)$ at time $t$ from expiration, of an option traded in the stock exchange, where $x$ is the price of the underlying stock. However, this example illustrates the fact that mathematics is really thought or understanding technology and not necessarily specific to given problems.

Physics is of course a great source of beautiful and important mathematics. The closest to the immediacy of physical processes that mathematics can come to, is probably via geometry. I want to close by going back to operations coming from geometry and physics.

In string theory, particles are represented by unparameterized loops (called strings). When moving they swipe out tubes. When interacting

69

they swipe out surfaces. For example a pair of pants corresponds to two loops interacting to become one. The surfaces have tubes at the ends that correspond to the incoming and outgoing strings. The analogue of Newton's law of inertia forces the surface to have "complex structure," e.g., a way of patching it up from pieces of the plane so that the gluings preserve the measure and orientation of angles. The surfaces can be composed by prescribing the gluing of an outgoing tube of one to an incoming of another. This is an operation on string interactions and also on pieces of the space parameterizing complex structures. In fact, it turns out that it can also be translated to an operation between certain kind of graphs, which one may think of as the backbones of the surfaces, in the sense that when we remove them, what remains from the surface can be smoothly contracted to a number of points. This way a geometric/physical problem can be translated to a combinatorial structure possessing also an algebraic operation. It is a thinking technique waiting to be thought.

Yiannis Vlassopoulos

Ivan Todorov
Bulgarian Academy
of Sciences

# Mathematical Physics

My parents were philologists (and so is my brother, Tzvetan), thus, as a child, I did not benefit from an early exposure to exact sciences. I inherited instead an interest in the history of science and a taste for communication between science and humanities.

It has been always fascinating to discover the same mathematical law governing apparently different phenomena. Instances of Felix Klein's pre-established harmony between mathematics (a game of pure thought) and natural phenomena seem even more impressive. The realization that numbers govern the sounds of music was a leading theme for Pythagoras and his school, the founders of mathematics as a science. Here is a more recent (and less popular) example of this type. After years of intensive study, following previous theoretical and experimental work of many physicists, Max Planck wrote in 1900 his formula for the energy distribution of black-body radiation that marked the beginning of quantum theory. Nobody seems to have noticed, however, that the small frequency (or high temperature) expansion of Planck's formula gives rise to the

Bernoulli numbers. Jacob Bernoulli (1654–1705) introduced his numbers in the context of probability theory. In the 19th century they were related to modular forms that are basic to analytic number theory. The integral coefficients in the Fourier expansion of such forms, which play a prominent role in the context of number theory, appear as multiplicities in the interpretation of statistical mechanics.

My young collaborator N. M. Nikolov and I pointed out (in a paper completed during my recent stay at the IHES, following a stimulating discussion with Maxim Kontsevitch) that the unique, normalized, modular form of weight four reproduces Planck's black-body energy distribution in conformally compactified space-time.

I also recall with pleasure a paper by Ya. Stanev and myself, initiated at the IHES (benefiting from the advice of another visitor, B. B. Venkov), in which the Galois group for the roots of unity was used to solve the Schwarz (finite monodromy) problem for the Knizhnik-Zamolodchikov equation, providing another link between number theory and, this time, conformal field theory models.

The legendary Alexandre Grothendieck (to whom the IHES owes much of its fame of the 1960s) explains in his "Recoltes et semailles" that a major stimulus for his abstract approach to algebraic geometry was the desire to find a common ground for the geometry of the continuum and the discrete «geometry of numbers». Alain Connes provides another unifying view of the discrete and the continuous—within noncommutative geometry. I vividly remember the series of seminar talks of Dirk Kreimer on his Hopf algebra approach to renormalization when the discussion (often continuing at lunch) revealed a close link with the Hopf algebra Connes and Moscovici had introduced, studying the transverse index theorem in non-commutative geometry. The visionary talk of Pierre Cartier at the 40th anniversary of the IHES, in which he dreamed of a unification of the ideas of Grothendieck, Connes-Kreimer and Kontsevitch, seems nowadays closer to reality than eight years ago.

Ivan Todorov

Anna Wienhard
University of Chicago

# In Praise
# of Tautology

A rigorous, correct mathematical statement is true independent of the circumstances, so it is always a tautology. This might make it seem empty or boring.

Mathematical thinking is exploring and discovering the unknown tautology. Recovering the diversity, creating and finding inner structures of the tautological, which, when regarded from the outside, as a whole seems meaningless.

But it unfolds to a beautiful world in which many tautologies are waiting to be discovered and created, and structured to form a meaningful whole which will never be completed.

Anna Wienhard

Giovanni Landi
University of Trieste

# New Worlds

Rather that the opposite of false or fallacious, for Greeks truth is *a-letheia* (α-λήθεια) namely what is un-veiled, revealed, not-ignored.

Continuously searching for minimal traces that would allow us to open doors on new and splendid worlds that would be unveiled 'enfin', we slowly move though concealed realms. And the lucky travelers-seekers bring to light, unveil, fragments of 'truth', rewarded with instants of pure, solitary joy.

Giovanni Landi

# The Music Pavilion

Pierre Deligne
Institute for Advanced Study,
Princeton
Fields Medal
Crafoord Prize
Wolf Prize

The heart of the IHES was for me the construction with the big bay windows on the edge of the wood, a former music pavilion, with, at the time, a library on the left and a conference room on the right. There, every Tuesday afternoon took place the Bois-Marie seminar on algebraic geometry where Grothendieck edified the algebraic geometry which is now familiar to us.

Contrary to what I sometimes hear said, his aim was not maximal generality. The ideal was that theorems should be well understood, that the structure of their proof should consist of definitions, carefully chosen so that each geometric idea should shed its light afar. A key word was "dévissage," or "unscrewing."

I was stunned the day he proved the base change theorem for a proper morphism $f\colon g^*Rf_* \xrightarrow{\sim} Rf'_* g'^*$ starting from the case of a constant sheaf on a family of curves.

The lemmas followed one another, self-evident, and, by the end of an hour, the heart of the theorem was exposed.

That manner of proceeding has remained my ideal, rarely attained: that nothing should remain visible of the efforts it cost to reach an understanding.

Pierre Deligne

77

Claire Voisin
CNRS - IMJ - IHES
CNRS Silver medal

# Long Live the Whale

The practice of mathematics ostensibly arouses curiosity, and yet mathematicians cannot talk about their subject without provoking boredom or pain. Moreover, each one of us would be delighted to do the honours of his area and help people to understand why a particular conjecture seems of capital importance to him. But expressions such as "topological invariants" or "polynomial equations", "metric curvature" or "moduli spaces" being censored, we can only take refuge in timorous silence.

To break out of this silence I will attempt a comparison. The picture which comes to my mind is borrowed from the world of *Moby Dick*. Like the watery element, mathematics is above all coherent, even if the immensity of the subject prevents contact between the most far-flung areas. As in the marine world, the hidden part, the submarine part, is infinitely more important than the visible part. It is that which carries us, even if the structures that it hides have not yet emerged, and will only later on change their status to become accessible to our understanding.

As for the mathematician, whether it be Ishmael, Ahab or Queequeg, he deciphers as far as he can that which is interpretable on the surface, yet keeps an eye open for the white whale (more or less impressive, according to his temperament or luck) which leads him to embark on these particular waters.

Claire Voisin

Jean-Marc Deshouillers
University of Bordeaux

# What Does One Write on It?

A blackboard. Two men sitting silent throughout the scene; one nods his head, gets up, writes a formula on the board and sits down again; the other one frowns, gets up in his turn, goes to the board and modifies the formula…. Mathematicians—male or female—don't recognise themselves in this caricature of their activity, of their communication, their social life. I take the opportunity here to pay tribute to the contemporary authors, directors, scenographers who represent mathematical activity more realistically and sometimes with great subtlety (I am thinking in particular of David Auburn and his play *Proof*, in French *La Preuve*). Not only are mathematicians not silent, but in addition, communication between them cannot be reduced to the exchange of "formulas". However, if the blackboard is not an essential instrument in communication between mathe-

maticians (you can "do maths" in a bus, while strolling along, during a meal…), it is nevertheless an important element, present in conference rooms and offices.

What does one write on it?

*Mathematical discourse* is extremely codified; besides, it is the formal aspect that the expression first evokes. It is its reason for being: the role of formalism is (and should only be) to validate the discourse. But the discourse is significant. As sophisticated as the form may be, it doesn't create meaning any more than the synthesis of molecules, however complex, creates life. When mathematicians talk to each other, they're communicating about meaning; they are manipulating concepts, exchanging mental representations. If this is common practice nowadays, I can remember, it was at

the end of the sixties, one of the greatest mathematicians who, when asked a delicate question, began by scribbling a tiny figure in a corner of the blackboard before quickly rubbing it out and serving up a totally formalized answer.

During the *phase of creation,* monologue or dialogue, the blackboard transcends its role of tool of reflection and takes an active part in the creation: it *reflects* the graffito, inflects thought, corroborating the remark of the poet, which I quote from memory: "What I write forces me to think that I was far from thinking that I was thinking".

It is also in the *phase of presentation,* essentially oral, that the blackboard is an aid to discourse: one writes on it the articulations, it's an aid to memory (the text remains on it often for more than half an hour). Filling it up gives the tempo of

the talk. But above all the blackboard enables us to represent mental images: it is certainly this role of tracer, of revelator of the mathematical activity of creation which makes it an important element in mathematical iconography.

Will the blackboard be left aside in favour of new technologies? The present craze for these tools suggests that possibility. And yet—even when new technology is mastered—it cannot convey the dimension of craftsmanship, the personal, physical, human dimension which the blackboard brings to mathematical discourse. Are we at a crossroads where the blackboard is still our wax tablet, with all the possibilities of immediate and spontaneous representation of mathematical art?

Jean-Marc Deshouillers

Pierre Cartier
CNRS
University Denis Diderot

# Solidarity

A mathematician who travels abroad is never a tourist. In a normal situation, given the professional aspect, he finds himself in a convivial environment where certain values are tacitly shared, humanist values which characterise more or less the mathematical community: mathematicians are liberals, despite a few blunders as during the international congress in 1920 in Strasbourg, where the Germans were excluded! They were only reintegrated in 1928.... However, Soviet mathematicians under Stalin did not come up against the same ideological errors as the biologists, for example. We must say they were in an ambiguous situation. The regime needed good scientists for its nuclear and military industry and tended to look after them, and mathematicians were sufficiently removed from the direct applications of their work to be deemed inoffensive, a bit like musicians or chess players. It's true that this position, somewhat in retreat, can be a haven in extreme situations and conduces to the development of a moral and physical aptitude to abstract oneself from reality which borders on autism. One takes refuge in mathematics a bit like

Miguel Angel Estrella in prison playing sonatas in his mind. I remember, at the worst moments during the Algerian war, always having a maths book in my pack which I would read whenever I had a quiet quarter of an hour to myself.

A certain number of us have taken advantage of this ambiguous status of mathematics when participating in international congresses held in countries ruled by dictators. It was the case in Ceausescu's Romania, it was the case in Czechoslovakia after the Prague Spring, where I went on behalf of the Jean Hus, an association presided by Jean-Pierre Vernant, Jacques Derrida and Nathalie Roussarie, a relative of Adrien Douady. I took with me banned books (Plato's *Republic,* for example) and a considerable amount of money, the equivalent of ten thousand euros today. Piotr Uhl, one of the leaders of the resistance at the time, was a native of Prague and a man of the old school: after having asked me not to stand too close to the window because of police surveillance, as soon as I had given him the packet, he sat down at his typewriter and typed me a receipt! I swallowed it as soon as I had gone out of the door.

Laurent Schwarz, Jean-Louis Verdier, Marcel Berger, Alain Guichardet and I also went to Poland. The situation was different, just after Jaruzelski's putsch, the 13th of December 1981. Speaking of this, nowadays there isn't a right-minded Pole who doesn't do justice to Jaruzelski. Thanks to his military putsch, he could say to the Russians, "No need to invade Poland, I have the situation under control." We should remember that his parents were killed in Katyn and that, if he wore dark glasses, it's because his eyes were burned by the snow in Siberia where he spent ten years in deportation. Anyway, it was December '81 and the International Congress of Mathematics was to be held in Warsaw in '82! The classical dilemma: to go or not to go? To caution or to boycott? Hence our exploratory mission in February '82. We found a society under close surveillance, with the appropriate number of police indicators and glamorous fortune seekers at the bar of the one hotel reserved for foreigners. The day after we arrived, at an official meeting at the Science Academy in Copernicus Square, the president comes onto the stage. "I have invited you in the name of the Polish Mathematical Society, unfortunately it has been banned as well as all the other associations...." However there was still a buffet, so we started to talk. A student comes up: "Are you the French people? I have a mission for you...." And he takes a letter out of his pocket and holds it out to Laurent Schwartz, who puts on his glasses to read it.... I kick him. "What's the matter?" I kick him again and this time he understands and puts his glasses away. The letter contained a list of a hundred and fifty imprisoned mathematicians. "They" hoped we could take advantage of the international congress to get them out of gaol. The young man who had given us the letter bends towards me: "Don't you want to come to Wroclaw?" I exchanged glances with Jean-Louis Verdier, always so solid and magnificent, and we said yes. The French Embassy, for whom we were the only presentable French delegation at that time, did the necessary work, and that very evening we had plane tickets and our passports duly stamped with visas.

Usually, at eight o'clock in the morning, Wroclaw is full of people, lorries, smoking factory chimneys. This time, nothing. General strike, town lifeless, apart from the patrols. As for us, we had no instructions. But a signpost indicates

"Ratuz"—Rathaus, that is to say Town Hall in Polish German—and there, on the main square, we find several student cafés open. We go in and feel immediately that we are on the right track: instead of finding us a table, the bar tender lets us stand around waiting like everybody else. So we are among equals, therefore in a democracy-friendly environment.... Indeed, once we have sat down, a young man comes up to us: "Are you the

French people?" His friends join him, so it isn't a police provocation, and he takes us to a meeting with a leader of the resistance, a Jewish mathematician who had been imprisoned, then released, and who wanted to discuss whether the mathematics congress should take place or not. The official regime wanted it to take place at all costs:

an international mathematical congress is a sign of normality with a lot of "added value." Finally it took place after a short delay, in 1983. Out of the hundred and fifty mathematicians on our list, a hundred and forty nine had been released. There remained the hundred and fiftieth, whose release Christophe Soulé and I obtained at the end of the congress. When the whole thing was over and we got on the plane home, I felt an envelope on my seat as I sat down. I put it in my bag and didn't open it until we got to Paris. On it was written: "Thank you for everything," followed by all the big names of Solidarnosc.

At that time, Jean Dieudonné also intervened a lot. As he was rather right-wing and didn't make a secret of it, it was natural for him to protest against communist leaders, but he also went to Montevideo and succeeded in persuading the Uruguay Defence Minister, a hulking great brute, to release José-Luis Massera by telling him: "In a civilised country, you can't do that kind of thing."

It was the right argument: the notion of civilisation is indeed a decisive one. I am firmly convinced that mathematics is an integral part of civilization and moreover that without mathematics civilization wouldn't exist. Obviously in some ways maths is that almost autistic activity I have already mentioned which can be done with just a paper and pencil but that only corresponds to one particular phase; Roger Godement once said jokingly that we should be attached to the arts faculty. Fortunately it was only a joke…. Mathematicians don't have tools, but they are constantly creating tools which just as constantly escape them and pervade all fields, from geometry to astronomy. And when my granddaughters start to count, they take in this activity like a gift of civilization which is their natural heritage. Mathematics reaches apotheosis each time one does maths without realizing it, and the true ambition of a mathematician is that mathematics should belong to everyone.

Pierre Cartier

Ali Chamseddine
American University
of Beirut

# North-South

I have spent some twenty years working in Europe and the U.S. and the remaining part working in a developing country. I am constantly preoccupied with the question of what does it take for researchers in mathematical sciences working in developing countries to be able to contribute to the advancement of knowledge at the same level as their colleagues in the North. What has been observed over the years is that most of the talented people coming from developing countries do their graduate studies in Europe and the U.S., the majority of whom do not go back to their countries, thus contributing to the brain drain. Some of the highly motivated graduates do return, and within a few years their research comes to a halt. They are perfectly capable of training students at the undergraduate level; the talented students among these pursue their graduate studies outside, and this cycle cannot be broken.

Governments in developing countries, overwhelmed with problems such as poverty and disease, do not have the means, or the vision to invest in research. I keep wondering, how many geniuses born among the poor nations did not get any opportunity to develop, and whose talents were lost. It is now certain that all races, given the same opportunities, can contribute to the advancement of science.

Sponsoring research centers is an ancient idea that played a fundamental role in the advancement of mathematics and sciences. The establishment of the Library of Alexandria at the beginning of the 3rd Century BC during the reign of Ptolemy II of Egypt served as a beacon of knowledge for many centuries. It also served as a research center where mathematicians such as Euclid, Archimedes and Apollonius worked under its roof. Another major event in the East is the establishment of the House of Wisdom by the Caliph Harun al-Rashid in the 7th Century. Later his son Al-Mammon sponsored many mathematicians such as Al-Khawarizmi, the Musa Brothers and Thabit ibn Qurra who contributed not only to the translation of Greek mathematics, but also added their own inventions such as algebra. It is easy to imagine that without these critical decisions in history, civilization would not have reached this advanced state. In the 19th-century research centers such as the mathematics institute in Göttingen, the mathematics produced by Gauss, Riemann, Dirichlet,

Hilbert, Weyl and others led the way to marvelous developments. In the modern era it is now universally recognized that civilizations must invest in science and mathematics as a guaranteed way to progress.

It is also well documented that for talent to flourish it must be nurtured, otherwise it cannot flourish. Because of this, it is the moral duty of rich societies to help the underdeveloped countries to discover and train talented people and to give them the opportunity to contribute to the advancement of civilization. The Indian mathematician Ramanujan provides the best example of how easy it is for geniuses not to be fully appreciated. I have no doubt that many potentially exceptional persons in developing countries are lost to poverty and disease.

The IHES provides the best example of how advanced societies provide nearly perfect conditions for talented scientists and mathematicians to be at the frontiers of research. The needs of such dedicated people are modest, and it is easy to calculate that the returns of investments that governments pay in support of fundamental research are astronomical. If a fund is established where 0.1% of profits resulting from spin-offs of scientific ideas are taken, then this will support mathematicians and scientists for a long time. One can only hope that politicians are kept aware of this fact, and that investing in science, not only in the developed world, but also in the underdeveloped world is the best bet to keep moving forward.

Ali Chamseddine

Christophe Breuil
CNRS - IHES
Dargelos Prize

# Privilege

After the École Polytechnique and the University of Orsay, the IHES was the third French institution I frequented, in September 2002. It's a place not quite like any other. For the first time, my employer, the CNRS, had given me leave of absence. For the first time, I benefited from a huge privilege, an office of my own. And, above all, for the first time I was immersed in a totally cosmopolitan universe.

The IHES asks almost nothing of its visitors, neither teaching nor administrative tasks, not even the obligation to obtain mathematical results (at least in the short term). At most—sometimes—one is asked to write a report to select potential visitors or postdoctoral research fellows. It is conceived like an earthly bubble where one has total freedom to do research and freedom to think, either completely alone or in active collaboration with other visitors (or both alternately). It is a haven of peace, isolated from the torments of the outside world. That is how, for five years, thanks to the support of the Scientific Council of the IHES, Laurent Lafforgue and I were able to invite more than fifteen close collaborators for long periods and to set up several seminars jointly with other organisms of the Paris region. That's how, for five years, thanks to the comfort that a research position at the CNRS affords, I was also able to pursue, in complete serenity, my research on "the programme of Langlands $p$-adic", which, out of respect for the reader, I will not elaborate on here.

For the IHES is more than just an institute for research in mathematics, or theoretical physics or biology. It is also a place where Culture and the quest for Knowledge, in the wide sense (not only scientific), are all-important. That poor culture, henceforth so mistreated that to possess just a little of it is enough to make you hopelessly old fashioned. That painful knowledge which, more and more reticent to be displayed, appears all the more vain as it is reserved for a tiny elite.

I remember countless conversations, scientific or otherwise, in front of a backboard, around the lunch table or over coffee. Conversations which were always interesting, rarely easy, where subjects as diverse as philosophy, history, literature, politics, religion, education or music were discussed (not to mention scientific subjects, that goes without saying). Because even if most of the mathematicians, theoretical physicists or biologists that I met from around the world devote themselves body and soul to their research, they are also discrete but brilliant personalities, often very cultivated; reserved but open-minded, critical but generous.

The researchers—Russians, Chinese, Indians, Americans, Africans, Europeans—have moulded the IHES in their image: a place where the most advanced theoretical scientific research is carried out but also a place where a humanistic, tolerant and often even disenchanted vision of the world is developed; a vision where the discovery of knowledge constitutes the ultimate achievement.

In a few weeks, I shall be leaving for a long stay abroad. I shall take with me thousands of hours of reflection spent at the IHES, dozens of wrong tracks laboriously pursued, a few mathematical discoveries but also hundreds of brilliant conversations; in other words an inexhaustible treasure. I shall be taking too, and above all, the memory of all the researchers who, like me, consider the search for truth (scientific truth in this case) a privilege and one of the noblest human endeavours.

Christophe Breuil

$$\boxed{\left(c\text{-}Ind_{KZ}^{G}\,\sigma\right)/(T)}$$

$\underset{\underset{I}{\overset{I}{\bigcup}}}{\overset{I^{(1)}}{\sigma}}\chi$     $\chi^s := \chi(wgw)$

$\chi > \chi^s$   $(\sigma, \sigma')$   $\chi \neq \chi^s$   $\sigma^s$ l'image     $g \in I$

$w = \begin{pmatrix} 0 & 1 \\ 1 & 0 \end{pmatrix}$

---

$\underline{\Pi \ \text{irréductible}}$ :   $0 \neq \Pi' \subseteq \Pi$   sous $G$-représentation.

$0 \neq \underline{soc}_K(\Pi') \subseteq soc_b(\Pi) = soc_K(D_0)$

$\Rightarrow \Pi' \cap D_0 \neq 0$

comme $\begin{pmatrix} D_0 \ \text{est irréductible} \\ D_1 \end{pmatrix}$     $\Rightarrow \Pi' \cap D_0 = D_0 \Rightarrow soc_K(\Pi') = soc_K(\Pi)$

$\begin{pmatrix} \Pi' \cap D_0 \neq 0 \\ \uparrow \\ \Pi' \cap D_1 = (\Pi' \cap D_0)^{I^{(1)}} \\ = D_0 \Rightarrow soc_K(\Pi') = soc_K(\Pi) \\ \boxed{\Pi' = \Pi} \end{pmatrix}$

---

$\underline{\Pi \ \text{supersingulière}}$ :   Si $soc_b(D_0)$ a $2$ facteurs de J.H.

Calculons $\Pi^{I^{(1)}}$ contient un ... de Hecke (pour l'algèbre de Hecke de $I^{(1)}$)

M.F. Vignéras

Laurent Berger
IHES - CNRS - ENS-Lyon

# What Do Mathematicians Do?

There are several ways of doing mathematics. One of them, the one I want to describe, consists of attacking unresolved problems, those famous conjectures left aside by mathematicians who prefer to advance in their exploration of the unknown, leaving to others the task of checking the consequences of their theories. If the solution to these problems generally manages to get the theory moving and to enrich it, the motivation is usually more down to earth: the desire to be the first to find the solution to a problem, the pleasure of answering difficult questions, the satisfaction of realizing that several aspects in a field, which up till then seemed dissimilar, are in fact closely related.

To resolve a problem is generally to prove a result, that is to say, to explain with sufficient rigour and clarity why it is true. Here is an example: draw a triangle whose vertexes you will call A, B and C. Trace the mid perpendicular of the three sides, AB, BC and CA (the mid perpendicular of a segment is the line which is perpendicular to it and goes through the middle.) If you are careful enough, you will find that the three straight lines intersect at the same point. You will then be led to conjecture that the mid perpendiculars of a

triangle are concurrent (this should remind some of you of your high school maths classes). How can we prove this? You can do algebra: give names to the coordinates of the three vertexes, then calculate the equations of the mid perpendiculars, then check that the three equations have the same solution. It is a perfectly valid proof, if your calculations are detailed enough and your friends can redo them line by line and check that your conjecture is indeed true. That illustrates one aspect of mathematics: to prove is not to explain. Even if you follow the calculations line by line, you probably still won't understand why the result is true. Here is another proof which is more enlightening; it consists in trying to understand why the mid perpendiculars are special. The mid perpendicular of a segment AB is the sum of the points which are at the same distance from A and from B. In our triangle, the two mid perpendiculars AB and BC intersect at a point P which is equidistant from A and B (since it is on the first mid perpendicular) and equidistant from B and C (since it is on the second mid perpendicular) and therefore equidistant from C and A, so it is also on the third mid perpendicular, and therefore these three lines are

indeed concurrent. When we look at things from the right angle, they become much simpler!

Let's take now an example of a problem that no one knows how to solve. We start with a whole number $n$ superior or equal to 1. If $n$ is even, we divide it by 2, and if it is odd we replace it by $3n+1$. For example, if we start with 13, then we come successively to 40, 20, 10, 5, 16, 8, 4, 2, 1. The problem is to show that whatever the number was to begin with, we always end up with 1. According to Wikipedia: "This conjecture mobilised mathematicians to such an extent during the 60s, in the middle of the cold war, that a joke went around that this problem was part of a Soviet plot to slow down research in America." Good luck!

When you are faced with an unresolved problem, it's like finding yourself up against a wall. You start by banging on the wall in various places, in the hope of finding a weak spot, a place where the bricks don't fit well and you can get at the masonry. Often it resists, and so a long period of incertitude and determination begins—mathematicians are familiar with this—during which you try out the tools which you possess, you use methods that you know well and you try and understand what is going on. As this is a fairly frustrating procedure, it's helpful to work on several problems at the same time (often they shed light on each other), to discuss these questions with colleagues, and generally to think about other things to stop the brain from overheating. After a more or less long period (a few days, a few weeks, even a few months), one finally makes progress. For example, you have treated a particular case and it seems representative of the general case, or maybe you have separated the problem into several different cases and some of these

seem feasible, or maybe you haven't understood what's going on but you have made lots of calculations which seem to give the right answer. As soon as you have found a weak spot in the wall, you can attack it seriously, and the more you advance, the bigger the section of wall you can knock down.

We then move on to the second stage: we can see more or less what has to be done, and we have to write up our solution and check the details. Even if you have a general idea about how to prove a result, writing it down can reserve some surprises: you realize that a particular short-cut is in fact a trap, that a particular intermediate result is not strong enough, and you sometimes have the impression you're trying to plug holes in a flimsy balloon filled with water; as soon as you have mended one hole, it starts to leak somewhere else! And if all the holes are too big, then you have to go back to the beginning. But as soon as you have plugged all the holes, then you can calmly finish your work and produce a solution as elegant as possible to the problem you had set yourself. And the conjecture is proved!

That is how some mathematicians work. But what does the mathematical community do? The object of a mathematician's work is to understand mathematical phenomena and to synthesize the sum of knowledge that we have by unifying it, by finding the right way to look at things. The thousands of results that we prove every year serve to push back and mark out the boundaries of what is true. Then, stepping back, we can explain it in depth and integrate it into the permanent body of mathematical knowledge. To accomplish this work, everybody's contribution counts.

*Laurent Berger*

Matilde Lalin
University of Alberta

# Fractalitas

Mathematics is like a treasure hunt with some differences. To begin with, it never ends. Solving the clues may bring more joy than finding a treasure. Moreover, there are infinitely many treasures and each of them is, at best, a clue for another hunt! In fact, mathematics may be more appropriately described as a fractal treasure hunt!

Matilde Lalin

Jürgen Jost
Max-Planck-Institut
für Mathematik, Leipzig
Leibniz Prize

# Mathematics, Biology and Neurobiology: A Profound Interaction

The interaction between physics and mathematics has been very fertile during the history of these two scientific disciplines. Mathematics furnishes the conceptual foundations and the formal methods for physics, and physics is a very rich source of problems and inspiration for mathematics. Newtonian mechanics motivated the development of infinitesimal calculus and the variational calculation of Euler; Riemannian geometry constitutes the foundation of the theory of general relativity; and more recently string theory has provoked profound repercussions in pure mathematics. The other sciences benefit from well-known analytic, numerical and statistical methods and set a number of difficult problems for applied mathematics, but without the profound interaction that is apparent in the case of physics.

Today, signs indicate that that may change. Given the spectacular experimental advances and enormous quantities of new data available, biology and neurobiology find themselves in a situation where, on the one hand, traditional formal and conceptual methods are no longer

sufficient to understand the data, and on the other hand, for the first time, we have a realistic chance to understand living systems in all their complexity. For mathematics this represents a historic opportunity which could be extended to the totality of their diverse approaches. Algebraic methods are able to identify, systematise and analyse discrete structures such as the gene or the information, around which biological and cognitive systems are organised, and to discover the underlying structures using the enormous quantities of biological and neurobiological data which we have accumulated. Geometric concepts help to explore the spatial organisation of biolo-

gical systems and can establish multidimensional relationships in abstract spaces. Mathematical analysis is indispensable for studying, modelizing and simulating deterministic or stochastic dynamics of a cell or of a neuronal network.

The IHES has played a decisive role in the development of pure mathematics for fifty years. Its vast experience in nearly all areas of mathematics gives it a unique opportunity to make in-depth progress in the totally new direction which research on biological and cognitive systems has initiated.

Jürgen Jost

Henry Tuckwell
Max-Planck-Institut
für Mathematik, Leipzig

# Mathematics in the Biological Sciences

When I was a graduate student I was confronted with the following choice: continue with mathematical physics or go to the University of Chicago and do a doctorate in mathematical biology. In those days hardly anybody with a theoretical bent thought to do anything but the physical or mathematical sciences. However, apart from regular science courses, I had taken courses in philosophy and psychology and in the latter I'd had my interest aroused in neuroscience—neurons seemed more relevant to human behaviour than neutrons! I also had the book *Embodiments of Mind* by McCulloch and it seemed that studying the brain would be a fascinating thing to do. I had seen that Nicolas Rashevsky was editor of *Bulletin of Mathematical Biophysics,* which prompted me to apply to Chicago, where he had started the first mathematical biology program in the U.S. My mathematical physics professors in Australia wished me well, although I could tell they were sceptical—I was very surprised when they later turned to research in Theoretical Neurobiology

I thus ventured to Chicago where I found out that Rashevsky, an immigrant from Russia,had been removed for being politically incorrect. I took my first course in probability from Patrick Billingsley—he was a very jovial person and I thought he could hardly be a serious scholar, as he kept saying "OK?" at the end of every other sentence. Later I found out that he had written the fundamental book on weak convergence. I became interested in stochastic processes and became absorbed in the works of Ito, Kolmogorov and other Russian probabilists.

Mathematics in biology prior to about 1960 could be summarized in a few short paragraphs. The legacy of three centuries of theoretical physics was that nearly all linear problems could be solved, at least in principle. Unfortunately biology for the most part has not cooperated and requires non-elementary mathematics for most of its theories. The start of theoretical population biology was probably Malthus's essay in 1797 on exponential growth and his forecast of the end of the world by overpopulation. Branching processes had been introduced as early as 1873 by Galton and Watson. There had been significant advances in mathematical genetics in the early part of the 20th century—beginning with Wright and Fisher, as they put Darwinian ideas into mathematical

form. I read that Darwin regretted not having had a better grasp of mathematical ideas. Hardy, who scorned applied mathematics, became famous, probably to his chagrin, for the Hardy-Weinberg law in population genetics, which is regarded by some as the Newtonian first law of genetics. Diffusion processes (probabilistic) had been employed by Wright as early as 1930, although it was not until many years later that mathematicians like Ito, Kolmogorov and Feller made the subject rigorous. In a similar way Fisher had proposed his partial differential equation for gene spread as early as 1937, but reaction-diffusion systems were not studied much by mathematicians until much later. Other pioneering influential works in mathematical biology were the Lotka-Volterra equations,

Kermack and McKendrick's papers on epidemics in the 1920s, and Turing's model for morphogenesis in 1951. Schrödinger in 1944 also entered the fray with his popular lectures and book *What Is Life?*, where he proposed that genetic material was molecular. This is supposed to have guided Watson and Crick in their discovery of the role of DNA. In brain theory, the extremely talented mathematician Wiener had done pioneering work in cybernetics and the remarkable Von Neumann had also made contributions.

I taught a course in theoretical neurobiology at UCLA in 1980 and later Cambridge published a two-volume work based on the course. At that time I thought they would not get much use, but in the next 10 to 15 years they became widely used as textbooks in diverse departments throughout the world. While researching material for the course I found that Lapicque had devised a commonly used neuron model in 1907, which many people have since written about. Physicists took up the subject with enthusiasm after Hopfield presented his simple yet powerful neural network model in 1982. Today the number of neuroscience-orientated articles in mathematics, physics and many other journals is immense—so much so that it is hard to keep abreast of the literature, unlike when I was a graduate student. Computer scientists have also became involved in large numbers in studying the brain and artificial intelligence. Some researchers are modeling whole brains and including every neuron, just as others are modeling whole countries with every member of the population in epidemiological studies! According to a recent review by the American Association for the Advancement of Science, "Opportunities for quantitative thinking about biological systems are exploding". One can't help but agree and wonder how things will be at the end of the 21st century and beyond.

Henry Tuckwell

"We can only dream", but henceforth machines are a precious aid to the modelling of the form and matter of dreams. Give mathematicians a computer and they will soon use it as a grapple to accost even physics and life science.

It is also useful to place them in front of a piano. Two mathematicians had played together only once, ten years before. This time on the music stand there was a transcription for four hands of the Brandenburg Concertos. For two hours, after a taxing day of conferences, they deciphered it.

Katia Consani
Johns Hopkins University

# The Unravelers

The title of this book describes mathematicians with a simple and effective word: "The Unravelers". It is then left to the mathematicians a more refined elaboration on why they have been portrayed this way and why they have chosen to dedicate their professional lives and devote their intellectual energies to unraveling the mysteries of mathematics. One may wonder why, for example, the pleasure of pursuing studies in mathematics prevails in some human mind over the dedication to other sciences or arts. Of course, each mathematician has his/her own personal explanation and life experience. This choice sometimes roots back to the childhood or early youth: it can be attributed either to a stimulating intellectual atmosphere in the family or to a clever and unusual introduction to mathematics at school or else to a circle of friends with a common point of convergence in intellectual discussions...

In fact, one could also phrase these questions by asking why some of us are particularly receptive to the mathematical logic or even more suggestively phrased to ask why some people "listen" to mathematics with a sensitivity that is still rarely understood while it would be much more easily appreciated if the subject of such attention had been, for example, music. After all, there are deep connections linking mathematics and music, everybody knows it. One could date them back to the ancient Greece culture that was fascinated equally by harmony and numbers. It is quite frequent to meet mathematicians with a refined musical taste that is expressed either in the form of playing satisfactorily an instrument or in the form of a cultivated theoretical knowledge. Mathematics and music are intimately related: both sciences are concerned with structures and the creation of new patterns out of what is presented to us at the beginning.

In mathematics we invent theories that are abstract organizations of a multitude of cases. We speak of beauty in a mathematical argument when our effort to create new patterns is rewarded by the discovery of hidden new relations that we have never seen before and in which we recognize an intrinsic symmetry and possibly a new and unexpected link to a different branch. Artistic creativity is a common quality shared by

a composer and a mathematician, and it includes frequently unpredictable elements, moments of frustration and depression which alternate with moments of emotional excitement and happiness. This is the wild side of any creative work of art and science, and in mathematics it coexists with our search for order and rationality.

I think that the term "unravelers" will please many professional mathematicians because it implicitly refers to the mental tension and the experience of comprehension that accompanies us in our work, at every moment of mathematical creativity, whether we are alone at our desks or when we expose our mathematical results at a conference, or when we exchange, among colleagues, our latest viewpoint on a theory at a research institute.

Katia Consani

Oscar Lanford
ETH Zurich

# Long Live the Machine

It all started because I love machines. I am fascinated by the mere idea of having a robot at hand, so refined that one can give it orders of increasing complexity and it will carry out the required task, no matter how sophisticated it may be. Nowadays that seems perfectly ordinary, but in the beginning it was a miracle.

It was in 1963—I was still a student with a summer grant at the Lawrence Livermore Laboratory in California—that I had my first contact with computers. It wasn't easy: to begin with, only a few privileged people had access to the machines, and if you wanted to programme a machine you first had to write out your programme by hand, then give it to a lady who typed it out on a perforated card that was fed into the slot, etc. It was practically the first transistorised computer (before they worked with lamps, which functioned for a few hours then overheated and burned out...) which cost several millions of dollars and which only a few selected researchers had access to. I was totally obsessed, to the despair of my wife, who would see me go off at 8 in the morning and come back at 11 o'clock in the evening. However, these machines weren't adapted to the type of problems which interested me, and I finally gave up.

But my enthusiasm was still there and since the '70s, my mathematics has been related to computers. In fact, it was really here at the IHES that everything started, when Louis Michel ordered a particularly elaborate, programmable desktop computer. It was a revolution! Before, if you wanted to do computer studies seriously, you had to go through enormous computer centres where you were completely cut off from the machines themselves and where the delays were very long. So this little personal computer that you could consult directly using BASIC language was a miracle. I was convinced that the computer could be to the brain what the bicycle is to the body, a means to multiply the possibilities. And like a bicycle, a desktop computer remains on a human scale. I've always had a problem with gigantic, supercalculators. At the same time I was in Berkeley, where a new mathematics and computer studies building was going up. As often, where building is concerned, there was plenty of money and there was enough left over to buy a powerful interactive minicomputer. With its advantages and inconveniences, compared with the Hewlett-Packard of the IHES, it was the second cause which determined my future engagement.

At the beginning of the '70s, at the IHES, David Ruelle had developed some interesting ideas on turbulence in relation to the theory of dynamic systems, which gave rise to some very simple models that one could easily display on a computer. Of course, we could have done it without the computer, but what a lot of effort it saved us! The decisive moment took place at the end of the '70s with the work of Feigenbaum on universality and renormalization groups applied to the iteration of simple functions. He noticed that certain numbers from certain types of systems present striking resemblances. Studying this, he made formidable progress in the understanding of the theory of dynamic systems. All of that thanks to the computer and to the purely accidental discovery that two different systems produce similar numbers.

These days, everything has changed. I am again immersed in an exciting project in numerical computation close to the theory of dynamic systems, and I continue to marvel like a child at the progress of the machines and the software, always more powerful, more adaptable, easier to use, smaller, cheaper and which, in addition to doing purely scientific calculations, henceforth dominate, through the visual aspect, all areas of communication.

However, I cannot help regretting the passing of those heroic days. I have the impression that today the simple questions have been solved and that even if there is still gold to be found, one has to dig deeper…. I, myself, was lucky to have been caught up in this movement from the beginning, at a miraculous moment: the facts were there, they were simply waiting for us to be able to discover them.

Oscar Lanford

Jürg Fröhlich
ETH Zurich
Max-Planck Medal
Marcel Benoist Prize
Dannie Heineman Prize
Latsis Prize

# Arriving in Paradise

Usually, the life of a scientist is not an easy life, and to remain sane, it is good to be honest about this fact, and sometimes to reflect on what has gone well and what has not gone so well. Scientists who work on important, challenging problems all reach their limits, sooner or later, and are confronted with the experience of failure and of being beaten by colleagues, or rescued by collaborators. The question is how one copes with such experiences.

I was a pretty good student, because I could easily learn things by following my teachers' expositions at a blackboard or discussing with my fellow students, always willing to imagine that I might be able to do at least as well as they did. Most of the time, books do not catch my attention; having to work through exercises proposed by a teacher, in competition with many other students, usually represented a challenge I was eager to take on, and I was quite good at it. Learning new things from books works for me, in cases when I have to prepare lectures that force me to learn those new things and translate them into simple prose. Glancing through some of my old papers, I realize that I would never have learnt physics properly if I had never had to teach it.

I have missed a good many opportunities to work on certain interesting scientific problems, not because I could not see them, but because I was convinced that I wouldn't be able to solve them successfully. This is one reason why I always liked to collaborate with other people. When one gets stuck on a problem it is comforting to know that one's collaborators may also be stuck. It is comforting if they then find a way out of the deadlock, and it is challenging to try to do at least as well as one's collaborators. I find it difficult to endure the loneliness of a desk in an empty office. This is why my office is messy, and why I usually don't work in my office; I work when sitting at a table or standing in front of a blackboard and discussing with colleagues or students, or by writing letters to collaborators.

When I think back on all the different places where I have spent my professional life, arriving in Bures was like arriving in Paradise. First of all I had excellent colleagues, permanent professors like David Ruelle, or visitors like Tom Spencer, Henri Epstein, Krzysztof Gawedzki, Elliott Lieb, Barry Simon, David Brydges, Erhard Seiler, Bergfinnur Durhuus, Carlos Aragão de Carvalho,

Sergio Caracciolo, Alan Sokal (now a famous philosopher), and others. Never before and never afterwards was the atmosphere among colleagues so cheerful and friendly. What's more, most of the time, I could do what I liked. While I very much like *not* to be told what to do, such a situation of complete liberty would soon have become a nightmare for me if I had not had the good fortune of having absolutely marvelous collaborators, who kept me focused, almost during our entire stay at Bures.

The magicians at the Institute were the mathematicians. Pierre Deligne and Dennis Sullivan appeared to be able to pull fantastic mathematics out of their hats or out of thin air. Dennis was especially magical—I have seen him struggle with calculating the first derivative of an elementary function during lunch. But then he would create those visions of how to solve problems like the Feigenbaum problem, avoiding tedious calculations and instead relying on powerful geometrical-topological intuition. I admired his lectures, with his anti-Bourbaki style. Proofs occasionally consisted of rhythmic motions of his arms accompanied by unidentifiable noises. I also enjoyed the company of Alain Connes, who then still had time to discuss and to socialize. I attended his lectures introducing cyclic cohomology. Daniel Kastler patiently tried to explain the details of "les idées d'Alain" to me. He was Alain's prophet. My office at the Institute was located next to a little seminar. On Saturday mornings, when I tried to catch up with all the work that could not be done during the week, I would hear the sound of the lecturer in René Thom's seminar about applications of catastro-

phe theory. Lecturers in his seminar included mathematicians, physicists, biologists, tailors, linguists, philosophers, etc. "Anything appeared to go" in this somewhat postmodern circle. I hardly ever attended them. I believed in what Richard Feynman has expressed in the following sentence: "It's not philosophy we are after, but the behavior of real things."

Theoretical and mathematical physics have proved to be extraordinarily rich sources of good problems in mathematics, and mathematics was recognized already by Galileo Galilei as the language in which the Book of Nature is written. It is useful to institutionalize a dialogue between mathematicians and physicists, as it happens at the IHES. This dialogue has proved to be very useful for both mathematicians and physicists, including ones working there. (Ruelle and Sullivan have done important joint work, Connes and Damour have coauthored a paper, Connes has collaborated with Kreimer and is working with Chamseddine, etc.)

Mathematical physics is a very useful endeavor in two kinds of periods: first, during periods when progress in discovering new Laws of Nature is stagnating, so that one can take time to consolidate what previous generations have discovered and to study new "emergent phenomena" within already existing conceptual frameworks and theories. Such attempts tend to be intrinsically rather mathematical. Examples are celestial mechanics during the times of Hamilton and Poincaré, theories of the onset of turbulence, or renormalization theory, or concrete problems in quantum mechanics, statistical mechanics and transport theory, etc.

Second, mathematical physics is useful in times when deep conceptual problems are turning up which call for a radical rethinking of the foundations of physical theory and for new mathematical tools. The times when the mathematical formulations of electrodynamics and statistical mechanics were found, in the nineteenth century, represent an example. Maxwell and Boltzmann were mathematical physicists in every respect. The period from 1900 till 1926, when quantum mechanics and the relativity theories were discovered and atomism was vindicated, is an even better example. Einstein, Schrödinger, Dirac, Pauli, Born and others were mathematical physicists in the best sense. It is important to stress that our generation too is facing a deep conceptual problem—one that has confronted all generations of theorists for the past seventy years: the unification of quantum theory with general relativity in a new theory of space, time and matter. This problem has proved to be so difficult and thorny that real progress has been very slow, and a definitive solution is not yet in sight.

The success of a mathematical physicist is not measured in the number of theorems he or she has proved, but in his or her ability to comprehend the famous "behavior of real things" in a mathematically precise way. There is excellent mathematical physics involving insights that cannot be formulated as rigorous mathematics, at least not yet. But there is also a brand of mathematical physics that brings forward new theorems, as well as new insights into physical phenomena. I have tried to do both things, with varying success. I feel one has to try to adapt one's style and one's

methods of attack to the problems one wishes to solve. Prejudices concerning style, technique and methods tend to hamper progress. Generally speaking, I don't think it is a good idea to try to see all problems from a single "unifying" point of view or through the lens of only one mathematical theory.

Thus, I have worked on rather many different problems in various areas of theoretical physics. On a few occasions, I have even managed to provide some theoretical interpretation of phenomena observed in the laboratories of colleagues in experimental physics.

I use mathematics in an eclectic way and usually do not attempt to create new mathematical tools, but rather try to use existing ones. I

It appears to the outsider that many of the young string theorists hardly ever do anything but solve rather concrete and technically very demanding problems. It may seem that they share my opinion that good theories, such as a "quantum theory of space, time and matter", will emerge only from the solution of very many concrete, hard problems. The problem is that, most of the time, they don't appear to solve problems deeply rooted in physics, nor do they solve problems of relevance for progress in mathematics—however, some *do,* sometimes. The situation reminds me somewhat of the one during the later days of the old Bohr-Sommerfeld-Epstein quantum theory, before Heisenberg, Schrödinger and Dirac came along. So it may be that, at present, they are not creating more than the background sound out of which the sound of a flute or violin or harp will emerge, announcing a breakthrough in attempts to create a quantum (or whatever) theory of space, time and matter. We don't know when this will happen, but that background sound may help to keep us focused.

like to invent physical problems and then sketch how one might want to attack them mathematically. The period when one doesn't know how to proceed further in *really* solving them and when one has to try out lots of alternative strategies, get stuck and find a way out of a deadlock, when one has to do lots of calculations, and so on; this is the period that often loses me, unless I am helped by competent collaborators. Generally speaking, I don't believe in visions, but I like to develop views and opinions, including strong ones, and change them whenever compelled to do so. I tend to think that good theories, physical *and* mathematical ones, grow on the ground of really good, relevant, concrete problems and, once they are created, help in solving these problems.

To conclude, my experience as a physicist has been that there are lots of different ways to be successful in physics. We should have respect for people who make other choices that don't work for us but may work for them. What counts in the end is to remain intellectually honest and to be able to lead a decent and enjoyable life.

Jürg Fröhlich

Sylvie Paycha
University Blaise Pascal

# Go to the Board!

You can talk mathematics up to a point, then words are no longer enough. So you get out your pen; a formula scribbled on the tablecloth or on a metro ticket can help to sort out the problem and set the conversation rolling again. When the exchange becomes more stimulating and the formulas start to crowd each other out on the little scrap of paper, you get up, and out comes the chalk. It's the blackboard, formerly so threatening in one's schooldays, yet now so familiar, which comes to the rescue. There, you can get to grips with mathematics; you scribble, rub out, write over the top, and an as yet wobbly construction of ideas is elaborated of which only a few traces remain on the board. For the sponge does its job, threatening to wipe out the white characters which shyly creep across the board. After each vigorous wipe a big, damp stretch relentlessly covers those which survived the preceding wash.

Silence, calm; one puts the sponge down, looks at the formulas that remain. A few steps back to gauge, evaluate, admire a formula, a proposition, a proof still in the making.... A state of grace for the ephemeral scribblings still on the

blackboard. The slightest comment, a question, however tentative, may make the few surviving sentences, formulas or symbols disappear forever. So, you meditate....

Finally, the eyes which up till then were staring at the board glance at each other, shining, satisfied, sated; smiles form, a sigh is heard: that's what we were looking for. The time has come to pick up the pen left on the table to set down on a clean sheet of paper the traces of this mathematical exchange which remain on the blackboard: the last chalk-written vestiges of a vibrant exchange which will be wiped out forever when the room is cleaned the next day.

Sylvie Paycha

Dennis Sullivan
Stony Brook University
AMS Steele Prize
Élie Cartan Prize
Oswald Veblen Prize

# Lunch at IHES (1975–1995)

When I joined IHES in the fall of 1974, there were no duties—just the opportunity to think about mathematics. I needed more structure, so I chose the duty and fun of making sure the lunch was a useful source of scientific exchange.

So almost every day, I would arrive by lunchtime and participate in the discussion that continued through the meal, dessert, coffee and sometimes almost until teatime for a smaller hardcore group.

During these discussions, which were mostly in English, we could learn from Deligne and exchange problems and ideas. My ploy was to take some idea or puzzle that was currently on my mind and convert it into an idea or question that held some general interest for those nearby at the table. Often the example would be placed in the arena of a young post doc or even more senior visitor.

Another ploy was to be vaguely aware of what most people were doing or were concerned with. Often I didn't understand too much about the subject, just enough to notice if two people who hadn't been "introduced" mathematically speaking were actually working on the same thing. In this way, Cheeger's L2 theory was introduced to Goresky-MacPhersons intersection theory. Also Mostow rigidity theory was introduced to Bers's quasi-Fuchsian deformation.

There were numerous other "introductions" at more modest scales every season.

Sometimes I was the main beneficiary of such discussions. On one of these occasions, I remember discussing with a topologist Ron Stern a concrete calculation of distribution of distortion of a Moebius transformation on the line $(1/1+x^2)dx$. This explanation of a concrete example—required by him to understand my abstract statement—actually helped me by its concreteness and explicitness. This led to a proof that Kleinian groups didn't have deformations on the limit set (1978).

It was also important to have long narrow tables. This way, we could have quite technical conversations if need be with nearest neighbours without being impolite by exclusivity to further away neighbours, who could have their own conversations.

At a round table such indiscretions are less possible because each person "sees more of the table" in front of him. Such details are very important for scientific dynamics.

Another detail was the flexible hotel system and flexible arrangement in the kitchen. A few of us could decide at the last minute whether or not some visitor or someone not "prévu" could go to lunch—assuming there was still food left.

I believe that the number of visitors now means that this is no longer the case, and that some of the senior professors who do not go to lunch regularly cannot decide to do so at the last minute. I hope this is temporary, because lunch dynamics could be affected by losing some of the energy of the masters, which during my years at IHES was crucial—Pierre Deligne, Alain Connes, Misha Gromov....

Dennis Sullivan

Jacques Tits
Collège de France
Wolf Prize
Abel Prize

# Happy Day at Bures-sur-Yvette in the Sixties

I leave the Gare du Luxembourg around 9 o'clock in the morning in the direction of Saint-Rémy-les-Chevreuse. In the train, as I was expecting, I meet a friend ready to engage in conversation, on a mathematical subject, naturally (a strange train where the discussions between travellers, which are always impassioned, are not about the latest television programme but about objects with curious names). At the station of Bures-sur-Yvette, other travellers join us and the discussion continues, even livelier, as we cross a wood, rather like the enchanted forest of Brocéliande towards this marvellous institute, a modern version of Rabelais's "Abbaye de Thélème", dreamed up by the magician Léon Motchane, seconded by the good fairy Annie Rolland.

On the way to my office, I stop for a moment to greet our two presiding spirits, who are always welcoming, and here I am sitting at my desk between a beautiful blackboard and an ample supply of chalk, which, as everyone knows, are the indispensable tools of a mathematician. My work progresses but suddenly I have doubts about the validity of certain arguments. Fortunately Deligne is not far away, and he soon dissipates my doubts (everyone here knows his generosity and takes advantage of it too often, perhaps).

Mathematics is very absorbing, but it makes you hungry and the lunch bell soon rings. With some difficulty I tear myself away from my precious papers and walk over to the cafeteria where thirty or so mathematicians, physicists, and members of staff are already sitting down, chatting animatedly about a variety of subjects. When the conversation turns to a hot topic in mathematics, often fascinating but difficult, I do my best to follow, sometimes successfully, sometimes not, and I am always filled with admiration and envy for "those who understand everything". The food is pleasant, the wine too, and mathematics (or physics, according to the day) lights up the whole scene.

After lunch most of us go to the seminar. For me this is a source of neverending enjoyment. If I am lucky enough to understand everything that the speaker says (let's say Grothendieck, to give

an idea) my happiness is complete. But even if it's not Grothendieck, I certainly wouldn't miss these discussions, where we witness the birth of new concepts, whose power we can only guess, and above all the exceptional moments when there is disagreement among the great experts: then the discussion is lively, voices rise higher and we can see looming up the spectre of the fatal contradiction, but everyone knows that the drama will not take place and that we shall invariably witness the miracle, which amazes all non-mathematicians who have seen such impassioned jousts: the verbal struggle ceases suddenly and one of the protagonists, with a burst of laughter, lays down his arms and admits he was "being stupid".

At four o'clock, tea is served and groups form in front of the blackboards to clarify a few last details.

Later, in the train going home, these "mathematical events" still fuel lively conversations; they mingle with the conversations of the students on their way home from the university, discussing the days' classes. I am struck by their seriousness; here again I have a lot to learn.

At last here I am back in Paris and happy to see my wife, who has spent the day at the National Library (in the rue de Richelieu, where it was so pleasant to work). She too is pleased because she has found the books and documents she was looking for. We exchange our impressions of a full day. The prospect of a good dinner (my medieval scholar is also a "cordon bleu") makes us see life in a rosy light.

Jacques Tits

Wendy Lowen
CNRS - FWO

# The Flowers
# of Maths

What is a mathematician?
Not a scientist
Not an artist
But caught between the two
In a world of structures and truths
Creating the seeds
But proving the flowers
And vaguely hoping
That their scent will save the world

Wendy Lowen

Michael Berry
**Bristol University**
**Wolf Prize**

# The Arcane in the Mundane

How delightful to discover our abstractions clothing Nature's realities:

Singularities of smooth gradient maps in rainbows and tsunamis

The Laplace operator in oriental magic mirrors

Elliptic integrals in the polarization pattern of the clear blue sky

Geometry of twists and turns in quantum indistinguishability

Matrix degeneracies in overhead-projector transparencies

Gauss sums in the light beyond a humble diffraction grating

More fundamentally, we are repeatedly astonished to find, recently developed and quietly waiting, exactly the mathematics we need for physics: Riemannian geometry awaiting general relativity, matrices awaiting quantum physics, fibre bundles awaiting gauge theories of fundamental forces.... Should we be astonished? I think not. We are beings of finite intelligence in an infinite, inscrutable universe. In science, our individual intelligences cooperate, and we can understand more. But still, we are able to comprehend only those structures in the natural world that mirror our mental constructs. And at any stage of humanity's development, the most sophisticated constructs are those of our mathematics. Therefore our deepest penetration into the natural world is limited by our latest mathematics. As mathematics develops, more subtle features of the universe become accessible to our understanding. While our species survives, I see no end to this process—no "Theory of everything".

So, "the unreasonable effectiveness of mathematics in the natural sciences" is not unreasonable at all; on the contrary, it is inevitable. Not unreasonable, but wonderful!

Michael Berry

Nathalie Deruelle
CNRS

# Allegory

*"Space in itself, time in itself, are condemned to fade away like shadows, and only a kind of union of the two can preserve an independent reality."*

*Hermann Minkowski, 1908*

As usual after lunch when the weather is fine, X has discussed the partition of the day outside in the sun in the park, then has returned to the music room. As usual she opens her piano: soft pedal pressed down, she plays a few arpeggios to tame the silence, a few scales which rise and fall, crescendo, and finally a series of diminished chords which she allows to resonate, until they fade into silence. The music is there, ready to be discovered.

What will she work on this afternoon? Bach, Debussy that she had practiced the days before? Well, no: today she will come back to a piece which is more in her line! And she opens a partition, tattered from frequent use: Rachmaninoff's third concerto—the feared, the inevitable "Rach 3". Her whole life as a musician is hidden within these pages. The long theme which opens the work runs through her whole body and seems to define time and space; she has known it since adolescence; her first teacher presented it in class;

fascinated, she had asked him what it meant. "I don't really know", he had replied, humbly; and it seemed to her that she had been trying ever since to answer the question herself.... The cadenza of the first movement was her choice piece for auditions. With what determination she had spent hours rehearsing it! The scherzo of the second movement is covered with annotations, some of them written by the friend with whom she worked at the time, who now devotes himself to Vinteuil's sonata. They were both still beginners, and she remembers their impassioned arguments before they discovered that the clarinet's melody was simply the main theme inverted. After that she had had a go at the central development, but was obliged to give up: her hands were too small. Recently she has tried again to master these formidable tenth chords and has succeeded in getting round the obstacles, playing a few notes in arpeggios, admittedly...but she thinks that now she can allow herself a little...clumsiness in her recital, let's say! As for the transposed accords of the finale, for her they are the most brilliant, most triumphant ones to be found in music. For years some technical points eluded her and prevented her from fully expressing the spirit of them. Now she feels that she is close to the goal: but she concentrates a moment, quietens her thoughts and begins at the octave, one note with each hand, the first exposition of the theme.

One by one, students, professors and their guests have entered the lecture hall to listen to X, trying to convince them of the beauties of "Rach 3". For an hour they listen to the melodies which seem to stretch out like a river; the chords that resonate like a fanfare; the cadenzas seem to have been written for hands which are more than human. The audience appears to be caught up in these complex movements of the fingers which seem to fly across the keyboard, one over the other, twisting and untwisting; they feel the inner tension of the musician who has become anonymous, swept along by a maelstrom stronger than herself. This tumultuous torrent of sound gradually entices them along, snatches them from the grip of ordinary time—bearing them away astride a beam of light!—and they reach the spheres of the universal beauty of harmony: once again the miracle of music is accomplished. X plays the final chord. The sounds dissolve and as silence falls, time… absolute time…can start again. The audience applauds but questions are immediately fired; and for two more hours X explains the details of such and such a point. When at last everyone has gone, she will close the keyboard and, stepping outside, will see that night has fallen on the park.

Nathalie Deruelle

Of what use is an allegory if not to glide smoothly from one universe to another (just like an equation: a l[...] dressed in fine muslin, her eyes bound and holding scales = Justice)? Thus we can move away from the p[...] bathed in darkness to attempt a maritime parable, beginning with the wood of which ships are made.

*And of course, what makes the prestige of a liner, the glory of its crossings, is the quality, the fame of its passengers. They are indeed its reason for being; otherwise it would be a cargo boat. But even so, can one imagine a ship without its crew? Without the captain, the officers, the radio operator, the steward, the boatswain, the sailors?*

*Shutting one's eyes, one can see indeed that all the maritime literature from Melville to Conrad, whatever the philosophical or metaphysical preoccupations of the author, is built around life on board, which depends on the moods, the decisions, the actions of a discrete population of busy, united sailors, organised according to an efficient and inexorable hierarchy.*

*And for the rest of the crew, we should praise essentially their actions: without them the boat would be wrecked on the nearest rock or sandbank.*

*We no longer need an allegory: mathematics is done better if the tea is nice and hot, the food healthy, if each new visitor can find his office without getting lost, if the microphones in the lecture hall are switched on, if the computer network functions without a hitch, if the texts are transcribed in the TeX language, if the lawn is beautiful and the park full of flowers. When all these conditions are fulfilled,* savoir-faire *gives way to another dimension: charm.*

Minoru Wakimoto
University of Kyoto

# Correspondence

*We solicited the collaboration of Professor Wakimoto during his stay at the IHES, and a little later we received this text by email.*

While I was staying at the IHES in the early spring of 2006, many plants were still sleeping and it was too early for them to bloom. But the days at the IHES were very fruitful through stimulating discussions on quantum reduction and collaboration with Professor Kac. At the IHES, we found a good class of nilpotent elements, and named them "exceptional" nilpotent elements, and started the study of the representation theory of W-algebras associated to these nilpotent elements.

Also there is a nice bakery near l'Ormaille, (the residence where I was living) and I often bought my favourite baguettes there on the way back from the IHES.

*Eager to know more, we thanked him for his reply, and asked him another question: "What is quantic reduction?" He replied that he was a little busy just then, but we could continue our conversation in the* following weeks. When it arrived, his detailed explanation fully satisfied us.

There are two very important classes of infinite-dimensional Lie algebras and Lie superalgebras, "affine Lie algebras" and "superconformal algebras". These infinite-dimensional algebras are not only interesting in themselves, but are also important because of their applications in various areas of mathematics and mathematical physics. We are interested in the structure and representation theory of these algebras.

There exist remarkable differences between affine Lie algebras and superconformal algebras. For affine Lie algebras, there are important devices such as "non-degenerate invariant bilinear form" and "Weyl group", which play a very important role in their representation theory. For superconformal algebras, however, there are no such basically important tools, so the representation theory of superconformal algebras has been quite difficult.

In 1980s, the theory of "W-algebras" was developed by physicists, and also representations

Feigin-Frenkel's theory is the $W$-algebra associated to the principal nilpotent element. All of so far known important superconformal algebras are obtained by our theory. The theory of $W$-algebra via quantum reduction has an advantage that representations of an affine Lie superalgebra $W\ (G,f)$ of level $K$ naturally give rise to representations of the $W$-algebra $W\ (G,f)$ and we are now studying representation theory of superconformal algebras in this way.

*This reply whet our appetite, and we asked Professor Wakimoto what a researcher in mathematics or physics got out of a stay in a foreign institute, since as long as he has a blackboard, a piece of chalk and a computer, a mathematician is at home everywhere.*

In an institute, we can discuss directly and exchange ideas freely on the subjects which interest us. This would be the most important benefit. Another important reason is, I think, the "atmosphere" as you have just pointed out. While we are staying at an institute, we can share a place with excellent mathematicians from various kinds of fields. And then it sometimes happens that we get inspired, which leads to a new viewpoint, even in the case when our research areas do not seem to be so close.

*Realizing that Professor Wakimoto thought he had said enough about his craft, we finally questioned him about the "baguette" from the local bakery. What was the best part? The crispy crust? The soft inside? The smell of fresh bread? The way it crackles? Did he, like many French people, snap off the end to munch before he got home? He replied:*

of some superconformal algebras were studied by physicists. In 1990, B. Feigin and E. Frenkel found a homological method to construct W-algebras associated to affine Lie algebras, by using the chain complex obtained from tensoring the universal enveloping algebra and the space of charged fermions. This method is called the "quantization of the Drinfeld-Sokolov reduction" or simply "quantum reduction".

A few years ago, we extended Feigin-Frenkel's theory to any Lie superalgebra $G$ and any nilpotent element $f$ and gave a method to construct the W-algebra $W\ (G,f)$ of level $K$. In our viewpoint,

Yes, all of that! Its crust is very crispy, and inside it's soft with a slight salty fragrance! And my wife and I loved nibbling the bread hot from the oven on our way home.

Thank you very much for your interesting questions.

Minoru Wakimoto

Victor Kac
MIT

# Ilan

I have visited the IHES many times over the last 30 years and, with the exception of Grothendieck, have met all its members. However, I shall write neither about the inspiring discussions and collaborations, nor about the discoveries made in this great institution.

An old joke claims that God will never be able to get tenure at an academic institution since he had only one publication. Fortunately, God does have tenure, but even demigods need it, and here comes the IHES to provide for that. The present collection will undoubtedly focus much of its attention on these great men (a demigoddess is yet to come), and rightly so. But I shall not write about this, the most glorious side of the IHES.

Another great source of strength of the IHES is that it is meant not only for demigods. It provides a safe haven for a great variety of men and women, who for the most part have regular jobs elsewhere (with often a less ideal environment for work), but sometimes with no other place to go.

Ilan was one of these few men. He was very different from the rest of us. Always well dressed and funny, somewhat of a playboy, he was interested in a lot of things. One week he would give a talk on p-adic analysis, a few weeks later on the Archimedes method of counting the grains of sand, and yet another talk on an esoteric problem in number theory.

Once, in the IHES tea room, I called Ilan's attention to an article of a Moscow high school teacher on the problems used to eliminate candidates of Jewish origin at the entrance examinations to the Math Department of Moscow State University. Some time later I received his treatise with a thorough analysis of the problems and the time needed for their solution.

"I leave aside the moral side of the question," he writes, "my goal is to analyze the complexity of the problems involved." It took him two to three hours to find a solution for each problem, while for these same problems unwanted candidates had been given only a few minutes. Needless to say, all problems designed for the rest of the candidates had been solved by Ilan in a matter of seconds.

The urgency of the question "Should the perpetrators of the crime be invited to the IHES?" is fading away, as these people are nearing their retirement age. Also fading away is the postcard sent to me by Ilan with the photograph of the cafe on Place Contrescarpe, where he was writing his treatise.

Victor Kac

Mikhail Gromov
IHES
Wolf Prize
Balzan Prize
Kyoto Prize
AMS Steele Prize

# The Four Mysteries of the World

The first mystery in the world is that of the nature of physical laws. One believes the structure radiates from a single point, something uniquely distinguished by the highest imaginable degree of symmetry, the symmetry that gets diluted and dissipates as the universe unravels itself to a human observer.

The second mystery is that of life. The dissipated structural symmetry of physical matter evolves into another kind of structure, condenses to structural islands in the exponentially large sea of all imaginable outcomes.

The third mystery is the function of the brain. Here an accidentally developed, seemingly amorphous mass of organic matter is able, by following physically dictated pathways, to select a proper response from the double exponential space of (imaginary?) possibilities.

The only way to represent any of the three structures in a format comprehensible to the human mind (brain?) is to construct mathematical models.

Almost all we see in mathematics today has evolved under the influence of the first of the three mysteries. Mathematicians keep searching for the ultimate symmetry of the universe projected onto the human reasoning. But nothing like that has ever helped in elucidating the structures of life and of the mind (brain).

And here the fourth mystery comes in, the mystery of mathematical structure. Why and when does it appear, how can we model it, and how does the brain manage to create it out of the chaos of the external input?

Maybe we shall have a glimpse of the answer at the 500th anniversary of the IHES.

Mikhail Gromov

Étienne Ghys
CNRS - ENS Lyon

# Flashes

*University Library, Lille, January 1977.* Yves has just brought me "Thurston's notes", fresh from Princeton, and I am photocopying some hundred pages. I decide that that's what I want to do in maths. Excitement.

*Lille University, June 1978.* I am discussing my ideas with one of my heroes—Dennis Sullivan—whom I met an hour earlier. A door opens and someone invites Dennis to drink a glass of champagne. He answers: "No, thanks. I prefer to eat mathematics." Bedazzlement.

*Lille, the citadel, April 1979.* I am going round in circles like the bear in its cage that I see every day as I walk round the citadel, with one idea in mind: how to show that these wretched Jacobian determinants are bounded. Will I succeed one day? Doubts.

*Rio de Janeiro, bus 125, March 1981.* It's very hot, the bus speeds along Flamengo beach. I understand that the orbits of an Anosov flow are geodesic in their stable manifolds. Maybe I can exploit that to understand the Anosov flows in dimension 3? Agitation.

*New York, June 1983, 3 o'clock in the morning.* The beautiful construction I was so proud of has just crumbled. Three months for nothing—I am no longer proud—a stupid mistake—I haven't understood anything about cohomology groups. Depression.

*Mexico, December 1985, end of the Lefschetz colloquium.* The great topologist William Browder comes up to me: "I enjoyed your talk!" Wow!

*Geneva, February 1987.* André, Pierre and I decide to study together an article by Misha Gromov on hyperbolic groups which is said to be incomprehensible. Determination.

*In a plane from Penn State to Philadelphia, March 1991.* Two Anosov flows can be coupled in the same way as Fuchsian groups. Jubilation.

*ENS Lyon, December 1991.* Christopher, my first student, defends his thesis! Satisfaction.

*Chemin du Vallon, Sainte-Foy-les-Lyon, March 1992.* But of course, with the Schwarz derivative, one must be able to construct an invariant projective structure. Exaltation.

*Grenoble, Weil Lecture Hall, October 1995.* I am talking about dynamic systems before a lecture hall packed with secondary school teachers who listen to me fervently. Empathy.

*Lyon, January 1st 1998.* It's my turn to direct the lab. Will I be able to do it? Hesitation.

*IHES, September 1999.* I accept the job of director of mathematical publications at the IHES. Responsibility.

*ENS Lyon, December 1999.* Sorin, my seventh student, defends his thesis on rigid structures. Happiness.

*ENS Lyon, July 2001.* The colloquium AMS-SMF has just ended. Six hundred participants. All the organisation undertaken in good spirits by the whole lab. What a lab! Pride.

*Gaeta, Italy, 2003.* Misha Gromov has just given a talk on fundamental groups of algebraic manifolds. I am enchanted. I enjoyed it from beginning to end. Admiration.

*Ostend, June 1st 2005, four o'clock in the morning.* Modular nodules are born by Lorenz's geometric model. It remains to be proved! Later! Certainty.

*Madrid, August 24th 2006, 11h44.* In one minute I shall begin my conference at the International Congress. Anxiety.

*Blois, July 22nd 2007.* With thirteen other mathematicians I have just spent a week reading and commenting on the nineteenth-century texts from which came the proof of the theorem of uniformization in 1907. A whole chain of mathematicians, across several centuries. Continuity.

*ENS Lyon, September 3rd 2007.* Pierre defends his Masters Dissertation. He will be my sixteenth student. Optimism.

Étienne Ghys

David Eisenbud
MSRI Berkeley

# State of Grace

In 1975–76 I came to Paris for my first extended stay abroad, a year at the IHES. The whole experience was extraordinary! We came by boat—that really dates it—my wife Monika napping on the deck, expecting our first child. It was my first experience trying to make friends and live in a language other than English. I learned about a field distant from my previous work—singularities—and proved some theorems of which I'm still proud.

But these were not the only major effects of my visit. It was perhaps the opportunity to be close to some of the world's greatest and most interesting mathematicians, spending time there, in a relaxed environment that was most important. Mumford was there, and many others, active and accessible. I got to know Nico Kuiper, Bernard Teissier, Norbert O'Campo, Lê Dung Trang. But the special atmosphere at the Institute, the lunches together, the common experiences, were of fundamental importance to me. Strolling around the grounds thirty years later took me back to...

Pierre Deligne, just a few years my senior but long since a legend, was at that time a "permanent" member. I had done a little work that I thought might be interesting to him, and of course I explained it—or tried to. He saw at once that there was a little problem, a finiteness assumption that was not justified. The proof survived this revision, but the rapidity of his understanding and his ability to explain—anything, seemingly—at the perfect level for whoever was listening in one-to-one conversation impressed me deeply.

Deligne's tutelage was not confined to mathematics. He took me bicycling on trails in the hills across the valley, my first experience with mountain biking, before I even knew the name. He also taught me a lesson about mathematical society: I addressed him politely as "vous", but he explained that all mathematicians in France "se tutoient"—use the familiar form of "you" to each other—not because they were all laid-back, groovy people, but because they were "in principle»" all former students of the École Normale, the school that has produced so many great French mathematicians since its founding by Napoleon.

René Thom's famous seminar was still in action, and I learned an important lesson at its first session that year. Thom introduced the subject for the seminar, saying we would study consequences of such and such a theorem, proved in the seminar the year before. A brave soul in the audience raised his hand and proposed a counterexample. There was discussion, and after a while there was general agreement that the counterexample was correct. "Now," said Thom, in no way perturbed, "let us continue with the consequences of the Theorem." It was the big picture that counted.

During the year that I was there, Quillen and Suslin proved "Serre's Conjecture" (he insisted, correctly, that he"d only stated a "problem", but the name was universally used.) It was a problem that was discussed a lot by Swan, Kaplansky and their students around Chicago; I was asked to present the proof at the IHES—with Serre, among others, in the audience. I'm not sure I've ever been as nervous about a talk! During it, Serre raised his hand and pointed out a little flaw in the argument as I was giving it. I managed to repair it on the spot, but my friends in the audience told me afterwards that I seemed as worried as if it had been my own proof.

Twenty years after my year at the IHES, I became Director of the Mathematical Sciences Research Institute in Berkeley (MSRI), so I've had a lot of time and reason to think about the experiences that Institutes provide, and a lot of visiting young people to watch. I know now that my experience at the IHES was not so unusual in its impact. Time to think, the presence of great mathematicians, an atmosphere of mutual respect and contact—these are things Institutes provide, things that influence young people (and others too) very deeply. I am grateful to the IHES for these gifts.

David Eisenbud

# Violins

Mathematicians are like violin makers. They inherit a craftsmanship accumulated through centuries and their apparent modesty cannot hide the pride which this knowledge inspires in them. Part of the heritage has been lost: Fermat's proof looks like the secret of Cremona varnish.

From time to time, the virtuosos of science come and consult mathematicians to ask for help tuning their theories so that they ring true and harmonious.

*Christophe Soulé*

Matilde Marcolli
Max-Planck-Institut
für Mathematik, Bonn

# Mathematics as Culture and Knowledge

Mathematics is an intellectual activity, arguably one of the most sophisticated ever produced by human civilization. Hermann Hesse sketched a portrait of the activities of mathematicians through the metaphor of the Glass Bead Game. That was perhaps the best literary attempt at catching a glimpse of some inner workings of the society of mathematicians. One does not blame a fictional work for inaccuracy, but it is indeed a difficult task to say something meaningful about what it is like to do mathematics.

There are quite a few mathematicians who entertain a Platonist view of mathematics. By this I mean essentially the belief that mathematical objects and constructs enjoy some kind of existence in a "world of ideas", independent of the human mind. As in the case of the mythical heaven, the proposers of such beliefs tend to be quite vague on the location and consistency of this extramental Platonic world. One reason that is often produced in support of the Platonic standpoint is the effectiveness of mathematics in modelling the physical world. No doubt that Kepler's laws would eventually be observed and understood by any technological intelligence living on a planet bound by gravity to revolve around a star (but would such a discovery follow the course we know, had the planet been revolving around two stars?). However, one can hardly make an equally strong case in support of other very beautiful but far more abstract branches of mathematics.

If no one would perhaps doubt that any sufficiently evolved extraterrestrial intelligence would understand the notion of prime number, there is far less compelling evidence that they would have our same notions of derived categories or *shtukas*. Recent years have gotten us used to more and more sophisticated mathematics being recruited to account for increasingly complicated models of high-energy physics. This sort of evidence notwithstanding, I personally remain extremely sceptical about the Platonist hypothesis.

Our brains developed over millions of years of evolution by natural selection. The capacity to produce mathematics has an obvious evolutionary advantage in as much as it is the key to the development of a scientific and technological civilization. The prominent position that this spe-

cies of apes has acquired, compared to the other animal species of the planet, is a clear demonstration of the evolutionary advantage of the brain's capacity for scientific thought.

Other brains that are the product of a completely different evolutionary process in an entirely different environment might also achieve the same goal of a technological civilization while producing a substantially different kind of mathematics from the one we know. Not entirely disjoint,

surely (prime numbers), but with possibly a vast symmetric difference. The existence of extraterrestrial intelligence is purely hypothetical. Sagan and Shklovskii speculated beautifully about it in the seventies and I will leave it at that, Platonism and all.

If mathematics (a large part of mathematics at least) is not a glimpse of the Platonic paradise but merely a by-product of our brains and evolutionary processes, it loses none of its beauty because

of that. It is all the more interesting because it is a part of human culture, and it moves along with and is influenced by the development of the rest of our civilization.

The mathematics that we know today is the result of a long and tortuous itinerary of cultural development. It is far from being a static edifice, however. Its continuous, rapid evolution can be seen easily by mentioning a couple of significant statistics. MathSciNet, the main source of reviews of mathematical publications, lists a total of 2,245,194 items, growing at the speed of 60,000 per year (and those listed by MathSciNet are just a selection of the total number of mathematical publications).

The first important step for anyone interested in working in mathematics is an awareness of the vastness of the landscape. One of the major risks, in my opinion, in mathematics and in many other fields of human knowledge, is that of being naive.

One does not improvise oneself a mathematician. Becoming one takes about ten years of intensive training and careful study. That's just in order to accumulate the minimal amount of knowledge and skills that are needed to understand what doing mathematics is about. To start to actually do something still requires a few further steps.

One, which is extremely difficult to acquire, and is a good sign of having achieved maturity as a professional mathematician, is the capacity to sniff out what is interesting. There are many things in mathematics one can do for the sake of doing them. Marcel Duchamp entitled one of his provocative sculptures "Classify the combs by the number of their teeth".

Truly interesting mathematics is not a technical exercise in classifying combs. Often what makes a mathematical result surprising and interesting lies in discovering unexpected connections: a way of relating results and constructions that were seemingly unrelated, recognizing a similarity of structures across apparently different phenomena. This requires knowledge. One has to be able to comfortably navigate the existent in order to be able to envision the nonexistent.

Being naive in mathematics has (with rare exceptions) the sole effect of digging oneself into an obscure corner of useless game-playing. Knowledge is what provides the crucial lighthouses and nautical charts that allow working mathematicians to navigate their way safely across rough waters.

There are widespread romantic mythologies about lonely geniuses who don't read but still generate beautiful theorems. These are based on largely fabricated anecdotal accounts. In truth, a long time spent reading and acquiring knowledge of present and past mathematics is essential to creating interesting future mathematics. Isolation only means the drying up of creative abilities.

Besides its effectiveness as a catalyst of invention, the transmission of knowledge through the written word is what makes us human. It is the key to the advancement of civilization. We read and learn because it is a great pleasure to do so, because we are human beings who care about being not an isolated fragment but a part of humanity as a whole. As in John Donne's famous poem, "No man is an island, entire of itself; every man is a piece of the continent, a part of the main."

Mathematics is interesting to an especially high degree among the achievements of humanity, because it has a universality that can provide us with a way to bridge and transcend the insignificant geographic and historic differences that split humankind. It is the common language our brains are hardwired to produce, the one that drives our scientific and technological development and is at the same time a deeply philosophical and artistic kind of endeavour.

Indeed, this is a peculiar aspect of mathematics that singles it out among the various fields of human knowledge. It functions simultaneously with the modes of the hard sciences and with those of the fine arts. Flights of the imagination, visual and poetic imageries and esthetical considerations drive the development of the field and live side by side with the most stringent rules of scientific rigor.

It is a pity that neuroscientists trying to understand how the brain develops mathematics generally tend to confuse mathematics with "num-

ber sense". The latter is a completely different faculty of the intellect, which is often completely disjoint from mathematics itself (there are plenty of examples of famous mathematicians who have no number sense at all). Mathematics is about the creation of structures and, in particular, numbers also exhibit interesting structures, but that's as far as the connection goes.

Understanding how mathematics is created in our brains will be a wonderful way to discover more of the functioning of the brain itself, since it provides a full spectrum of modes of operation of creativity and imagination as well as of manipulation of images and symbols, with a very precise and well-defined focus.

The final answer, if one is needed, to the question of why we do mathematics, is that we derive pleasure from doing so. It is an obvious by-product of evolution by natural selection that we derive pleasure from doing things that are also beneficial to the survival of our genes. Mathematics is beneficial to our species because of the applications it has to our science and technology, but that's not the reason why we enjoy doing mathematics. We do not think of the importance of practical applications when we enjoy creating new mathematics any more than we think of the importance of mixing our DNAs when we enjoy having sex.

Matilde Marcolli

Alessandra Carbone
INSERM
University Pierre et Marie Curie

# A Question of Time

The interface between mathematics and molecular biology, which for some years I tried to investigate, being at the institute, represented for me a novel and different intellectual activity. Mathematics, as I was doing it at that time, is based on no data, contrary to math-biology which requires mathematical thinking, where progress, methods and intuition depend strongly on the analysis of datasets. Theorems are replaced by a numerical verification of hypotheses over known datasets, and in fact, any progress in the understanding of evolution and coevolution does not seem within the reach of purely theoretical thinking. In some cases, probabilistic contributions have been demonstrated to be fundamental, as has been the case for the Lander-Waterman analysis for genome sequencing. The problem of establishing conditions ensuring a parallel reconstruction of genome sequences of multiple organisms is still open, and probabilistic modeling is asked to clarify such fundamental questions, which will lead to a different understanding of the biology of microbes and their environment. Genomics moves today towards integrating "environment" in its definition, and on this line of thinking, *homo sapiens* is understood through his interaction with millions of different microbes determining his environment. Understanding unicellular and multicellular organisms together with their environment constitutes the leap we are expecting from biology in the near future, and a mathematical formalization of appropriate spaces and measures might indicate a way towards the environmental classification of organisms.

Certain biological phenomena, once defined, turn out to be robust concepts. To unravel and state them properly is a challenge. Guessing at correct definitions and biological hypotheses requires mathematical thinking, and the screening of the various data configurations requires sophisticated rigour and a critical attitude. Claims are proven numerically, with an interlacing between statistics and combinatorics, and possibly without being supported by biological hypotheses. Algorithms play a fundamental role in data analysis, and heuristic approaches and design of algorithms for fast, exhaustive searches bring us to face the theoretical limits of computational feasibility. The detection of biological signals by playing with computational processes is a challenge. The interplay between statistics and combinatorics needed to answer biological questions, and which is strongly linked to an insightful understanding of evolution, should be developed. The way to do it is hard to see and will probably take many years.

Whether or not the interface between mathematics and molecular biology ever will cohabit with mathematics (and physics) at the institute, only time will say. At the moment, they might look worlds apart.

Alessandra Carbone

Jean-François Méla
University Paris 13

# "The Times, They Are A-changing"

When I was writing my thesis at what was not yet called the Science Faculty of Orsay, there was a seminar there on harmonic analysis (originally the Kahane-Malliavin seminar) which the participants still remember with emotion and pleasure. One would meet there the top names of the world of "fine analysis": Zygmund, Carleson, Rudin, Stein, Katznelson...). A passionate and cosmopolitan group of young people collected there, bringing their contributions. There was an air of adventure: each week brought its share of discoveries and strange problems. One nostalgic day, Yves Meyer, one of the bright lights of the seminar, confided in me, hyperbolically: "It was so fantastic that we should all have died at the end."

In those happy days, it wouldn't have occurred to us to cross the river to go to the monastery opposite which was the temple of a prestigious brotherhood: algebraic geometry and the disciples of Grothendieck. The old-fashioned aspect of "fine analysis" and the "entomological" approach of mathematical beings would have been gently mocked (fractals were not yet in fashion…).

At that time the geography of mathematics was interesting but a bit of a patchwork. The notion of laboratory or department in the Anglo-Saxon sense was still in limbo. Orsay was a pioneer in the matter. Often any collective ambition was thought suspect. Wasn't that to encourage general mediocrity at the expense of individual excellence?

Ten or fifteen years later, the beautiful young people in mathematics had aged. In 1986, in France, half of university professors were between 40 and 47 years old. That year there were only sixteen senior lecturer posts available. Even the very modest place of mathematics in the CNRS was contested. In addition, as far as money was concerned, we were only allotted pencils and erasers. Thus France made only a miserly contribution to the international congress in Berkeley where our diplomatic corps was singularly absent.

We discovered that the world had changed and that we would have to be able to convince our contemporaries (as our American colleagues had done) that mathematics—all mathematics—is a "strategic resource for the future." A cruel dilemma for a community more preoccupied by its theorems than by its economic and social role. I was privileged at the time to belong to a "band of pals", fully aware of what was at stake, who had taken over the French Mathematical Society and urged me to become president to sound the wake-up call. The visionary of the gang was Jean-Pierre Bourguignon. I am sure I will offend his modesty by saying that, but too bad! His influence has not waned since, including on the European level.

In 1987 the colloquium "Mathématiques à Venir" held at the École Polytechnique precipitated a change on the part of the decision-makers in favour of mathematics, which up till then had been absent from important scientific programmes. There was also a realization in the community of the importance of a collective, European and planetary organisation of which the IHES has today become one of the leading lights, open to the plurality of mathematics. But that is another story.

Jean-François Méla

Jean-Pierre Bourguignon
CNRS - IHES

# The Vision

## From Which This Book Has Arisen

There are no foggy mists as there are no algebras
Which can resist, in the depths of numbers or skies,
The calm deep gaze of eyes;
I looked at this wall at first confused and vague
Where everything seemed to float as on a wave
Where all seemed vapor, vertigo, illusion,
And beneath my pensive eye, the strange vision
Became less misty and more clear
As my eye was less troubled and more sure.

The discovery of this text by Victor Hugo which opens the "Légende des Siècles" at first amazed me; I found it so pertinent…to use imagery, couched in such superb language, to describe the activity of a mathematician embarking on a work of some importance. Then surprise turned to a feeling of the obvious: any wide-reaching project must stem from a preliminary outline, nurtured by a "vision" that perseverance alone could turn into a document worthy of being submitted to the eyes of others. More than that: the author, the unraveler, I would venture to say, after having worked on his text over and over again, through dint of studying, acquires a conviction, a capacity of expression, in a word a momentum he would never have imagined possessing. Isn't this conquest through effort the most precious lesson that studying can give us?

However, there is no reason to limit this approach to the production of texts: it is also largely relevant for many other human enterprises of any importance. Léon Motchane was also moved by a vision when, against all odds, he conceived and developed the IHES, an institution, which

by its originality, was totally improbable in the France of the 1950s. He believed in its destiny, and certainly forced it at times, swept along by the inspiration which dissipates all illusions. The relay was taken up and the Institute continues its road, but it needs permanently to recreate its destiny: that is the law of exception, to have nothing to copy.

To come back to Hugo's epic poem, couldn't we dream of another outcome which would put mathematics at the heart of the "Légende des Siècles"? Nurtured by different cultures and transcending them, mathematics gives them a new dimension. Their adventure, constantly renewed, bounds on from challenge to challenge, whether they spring from its own territory or well up from exterior sources, brought to light by practitioners of other sciences, engineers, or simply curious scholars who, faced with an interrogation, are not satisfied with negligence.

Don't let us cherish illusions. Even if a few giants like Old Hugo can claim consecration in the universal memory through their quest for beauty fulfilled and their dazzling conquest of a few certainties, we still have the exulting task of executing our work as craftsmen. We must make sure we have given ourselves the means to continue the chain and transmit, to a few apprentices, the tricks we have patiently learned at times when our eyes, more sure, were no longer troubled.

Jean-Pierre Bourguignon

Denis Auroux

CNRS - MIT

# Taming Mathematics

When a stranger asks me what I do, and I declare myself a mathematician, the reaction often betrays a certain anxiety ("math never was my thing") but also some curiosity ("what kind of research does one do in math?"). Taming mathematics requires constantly renewed efforts, and at the beginning of a mathematician's career it is certainly necessary for curiosity to overcome anxiety.

I remember my first contacts with mathematical research: the apprehension when faced with concepts that seemed forever out of reach, the frustration of not understanding. Curiosity pushing me to return unendingly to those seminars, starting each week with the ritual mantra "let $X$ be a Calabi-Yau manifold" followed by incomprehensible and terrifying incantations.

Today, thanks to all those who have helped me to tame mathematics (they will recognize themselves), I am no longer afraid. I am now a true mathematician, and my talks also start with "let $X$ be a Calabi-Yau manifold". What follows this little sentence is undoubtedly terrifying for

at least one person in the audience. But I hope that for that person, my explanations will help him or her to tame a tiny fragment of mathematics. Not necessarily to understand it, but to be less afraid, a necessary prelude to understanding and progress on the long road of mathematical knowledge.

Denis Auroux

175

Alexandr Usnich
École normale supérieure

# Three Lights

## Lunch

One day I had lunch with Shigefumi Mori, the great Japanese mathematician. On no account would I have missed the opportunity of talking to him, but I wanted to avoid questions of mathematics. After a little reflection or through reading, I could always answer the questions myself. However, I deliberately chose to adopt the attitude of a disciple, longing to ask the most general existential questions. And so, just before attacking the raw fish, I asked him what advice he could give me, in life or in mathematics.

And like in a Buddhist legend, the reply was simple yet full of meaning: "One must do what one profoundly desires to do." It was crystal clear. One does not learn: one convinces oneself of the truth of what one has always known. The simplest words are the most inspired.

## The Unknown

There are few things more democratic than the unknown. Faced with the unknown, no judgement is better than any other; all are subjected to its implacable law. The unknown in science has a human dimension: you advance and the shadows disperse. It's the "soft" version of the unknown. A far reach from the unknown of doctors and soldiers who look death in the face, who meet the eyes of those who see death, the unbearable unknown. The unknown instigates the quest which in turn kills the unknown. The irrepressible need to know as if we were afraid of not knowing.

## The Unforeseen

The unforeseen which paints life in vivid colours, opening the doors to new possibilities. Which guides our lives in as much as we allow it to. Ignorance of what will happen can bring about disappointment: why are things not as we imagine them? Perhaps we don't think enough, perhaps if we were diligent enough we could see in advance what will happen. Wisdom leans on the past, stretches to the future and unveils the unforeseen.

Alexandr Usnich

Maxim Kontsevitch
IHES
Fields Medal
Crafoord Prize
Iagolnitzer Prize

# Beyond Numbers

Very often a mathematician considers his colleague from a different domain with disdain—what
kind of a perverse joy can this guy find in his un-
motivated and plainly boring subject? I've tried to
learn the hidden beauty in various things, but still
for many areas the source of interest is for me a
complete mystery.

My theory is that too often people project their
human weaknesses/properties onto their mathematical activity.

There are obvious examples on the surface:
for instance, the idea of a classification of some
objects is an incarnation of collector instincts,
the search for maximal values is another form of
greed, computability/decidability comes from the
desire of a total control.

Fascination with iterations is similar to the
hypnotism of rhythmic music. Of course, the
classification of some kinds of objects could be
very useful in the analysis of more complicated
structures, or it could just be memorized in simple cases.

The knowledge of the exact maximum or an upper bound of some quantity depending on parameters gives an idea about the range of its possible values. A theoretical computability can be in fact practical for computer experiments. Still, for me the motivation is mostly the desire to understand the hidden machinery in a striking concrete example, around which one can build formalisms.

If one tries to go further towards the "dehumanization" of mathematics, a natural next step would be to consider the real numbers (which emanate from the basic properties of the physical world) as just another complicated non-algebraically closed field.In some sense it is true; the complex numbers are much more beautiful.

But in another sense the real numbers are truly fundamental as they incarnate the idea of a bound, of a control of abstract algebraic structures. In a deep sense we are all geometers.

Maxim Kontsevitch

# Photographs

*The photographs were all taken by Jean-François Dars between January 2006 and September 2007 at the IHES with the exception of those noted.*

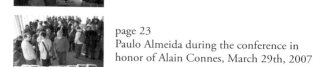
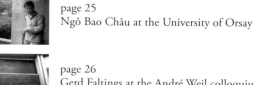

page 134
Valérie Landais

page 134
Cécile Cheikhchoukh

page 134
Thibault Damour and Marie-Claude Vergne

page 134
Malvina Dussart

page 135
Jean-Pierre Bourguignon

page 135
Marcel Berger

page 136
Dominique Guiet and Geneviève Miénandi

page 136
Christine Bontemps

page 136
Hermann Baliziaux

page 137
Catherine Nguyen

page 137
Marc Monier and Chanel

page 137
Joanna Jammes, Aurélie Brest, Caroline Beausire,
Catherine Nguyen, Emmanuel Hermand,
Nathalie Carré, Hélène Wilkinson, Hermann
Baliziaux, Christine Bontemps, Régine Lepori,
Laurence Beaupparain, Filomena Seabra, Marc
Monier, and Luis Afonso

page 138
Patrick Gourdon and Jennifer Yobusa

# Thank You for Everything

Jean-François Dars, Annick Lesne, and Anne Papillault are honoured to belong to the CNRS, which maintains, like glowing embers beneath the ash, the essential spirit of its founders and, as years go by, of its most talented directors: their unconditional generosity, although those who benefit from it have indeed the mission, implicit in the social contract, to transform their ideas into real or realisable projects.

They thank the IHES for its qualities as melting pot, crossroads and outpost of fundamental research.

Two out of the three of them, as far removed as possible from the mathematical universe, have thus just grazed the surface. They have constantly had the impression of sharing the paradoxical experience of mermaids, earthy beings in a marine world, floating along with ordinarily extraordinary fellow creatures.

From this stay, during which they were able to contemplate the astonishing qualities and banal defects of those who people the world of calculations and hypotheses, they have retained one certainty: that whatever one can say about them is also valid for the whole of humanity, the only noticeable differences being the degrees of intensity.

They have also experienced their time here as a privileged moment similar to a journey in time where one could chose one's objective: for example, without disturbing them, to watch Lorenzo Lotto paint or Robert Schumann compose.

# Index